D1747392

Interfaces of Ceramic-Matrix Composites

Interfaces of Ceramic-Matrix Composites

Design, Characterization and Damage Effects

Longbiao Li

WILEY-VCH

Author

Prof. Longbiao Li
Nanjing University of Aeronautics and Astronaut
College of Civil Aviation
No. 29 Yudao St.
210016 Nanjing
China

■ All books published by **WILEY-VCH** are carefully produced. Nevertheless, authors, editors, and publisher do not warrant the information contained in these books, including this book, to be free of errors. Readers are advised to keep in mind that statements, data, illustrations, procedural details or other items may inadvertently be inaccurate.

Library of Congress Card No.:
applied for

British Library Cataloguing-in-Publication Data
A catalogue record for this book is available from the British Library.

Bibliographic information published by the Deutsche Nationalbibliothek
The Deutsche Nationalbibliothek lists this publication in the Deutsche Nationalbibliografie; detailed bibliographic data are available on the Internet at <http://dnb.d-nb.de>.

© 2020 WILEY-VCH GmbH, Boschstr. 12, 69469 Weinheim, Germany

All rights reserved (including those of translation into other languages). No part of this book may be reproduced in any form – by photoprinting, microfilm, or any other means – nor transmitted or translated into a machine language without written permission from the publishers. Registered names, trademarks, etc. used in this book, even when not specifically marked as such, are not to be considered unprotected by law.

Print ISBN: 978-3-527-34778-0
ePDF ISBN: 978-3-527-82803-6
ePub ISBN: 978-3-527-82805-0
oBook ISBN: 978-3-527-82804-3

Typesetting   SPi Global, Chennai, India
Printing and Binding   Markono Print Media Pte Ltd, Singapore
Printed on acid-free paper

10 9 8 7 6 5 4 3 2 1

To Shengning

Contents

Preface *xi*
Acknowledgments *xiii*

1 **Definition, Function, and Design of Interface in Ceramic-Matrix Composites** *1*
1.1 Introduction *1*
1.2 The Definition of Interface in Ceramic-Matrix Composites *2*
1.2.1 Non-oxide CMCs *3*
1.2.2 Oxide/Oxide CMCs *13*
1.3 The Function of Interface in Ceramic-Matrix Composites *18*
1.3.1 Effect of Interphase on Sliding Resistance *19*
1.3.2 Effect of Interphase on Thermal Misfit Stress *19*
1.4 The Design of Interface in Ceramic-Matrix Composites *20*
1.4.1 PyC Interphase *20*
1.4.2 BN Interphase *21*
1.5 Conclusion *21*
References *22*

2 **Interface Characterization of Ceramic-Matrix Composites** *29*
2.1 Introduction *29*
2.2 Effect of Interface Properties on Tensile and Fatigue Behavior of Ceramic-Matrix Composites *30*
2.2.1 Theoretical Analysis *31*
2.2.1.1 First Matrix Cracking Stress *31*
2.2.1.2 Matrix Cracking Density *31*
2.2.1.3 Fatigue Hysteresis-Based Damage Parameters *32*
2.2.2 Results and Discussion *33*
2.2.2.1 Effect of the Interface Properties on First Matrix Cracking Stress *33*
2.2.2.2 Effect of the Interface Properties on Matrix Cracking Density *36*
2.2.2.3 Effect of the Interface Properties on the Fatigue Hysteresis-Based Damage Parameters *39*
2.2.3 Experimental Comparisons *41*
2.2.3.1 First Matrix Cracking Stress *42*
2.2.3.2 Matrix Cracking Density *43*

2.2.3.3	Fatigue Hysteresis-Based Damage Parameters *46*
2.3	Effect of Pre-exposure on Tensile Damage and Fracture of Ceramic-Matrix Composites *51*
2.3.1	Theoretical Analysis *52*
2.3.1.1	Stress Analysis Considering Interface Oxidation and Fiber Failure *54*
2.3.1.2	Matrix Multicracking Considering Interface Oxidation *56*
2.3.1.3	Interface Debonding Considering Interface Oxidation *57*
2.3.1.4	Fiber Failure Considering Interface and Fiber Oxidation *58*
2.3.1.5	Tensile Stress–Strain Curves Considering Effect of Pre-exposure *59*
2.3.2	Results and Discussion *60*
2.3.2.1	Effect of Pre-exposure Temperature on Tensile and Damage Process *60*
2.3.2.2	Effect of Pre-exposure Time on Tensile and Damage Processes *61*
2.3.2.3	Effect of Interface Shear Stress on Tensile and Damage Processes *62*
2.3.2.4	Effect of Fiber Strength on Tensile and Damage Processes *64*
2.3.2.5	Effect of Fiber Weibull Modulus on Tensile and Damage Processes *65*
2.3.3	Experimental Comparisons *66*
2.4	Effect of Interface Properties on Lifetime of Ceramic-Matrix Composites *71*
2.4.1	Theoretical Analysis *73*
2.4.1.1	Life Prediction Model at Room Temperature *74*
2.4.1.2	Life Prediction Model at Elevated Temperatures in the Oxidative Environment *77*
2.4.2	Experimental Comparisons *79*
2.4.2.1	Life Prediction at Room Temperature *79*
2.4.2.2	Life Prediction at Elevated Temperature *89*
2.5	Conclusion *100*
	References *103*
3	**Interface Assessment of Ceramic-Matrix Composites** *109*
3.1	Introduction *109*
3.2	Relationships Between Interface Slip and Temperature Rising in CMCs *112*
3.2.1	Hysteresis Theories *112*
3.2.1.1	Case I *113*
3.2.1.2	Case II *115*
3.2.1.3	Case III *116*
3.2.1.4	Case IV *117*
3.2.2	Experimental Comparisons *118*
3.2.2.1	Unidirectional CMCs *118*
3.2.2.2	Cross-Ply CMCs *122*
3.2.2.3	2D CMCs *124*
3.3	Interface Assessment of CMCs from Hysteresis Loops *127*
3.3.1	Results and Discussion *128*
3.3.1.1	Unidirectional C/SiC Composite *129*
3.3.1.2	Cross-Ply C/SiC Composite *131*
3.3.1.3	2.5D C/SiC Composite *135*

3.3.2 Experimental Comparisons *138*
3.3.2.1 Unidirectional C/SiC Composite *138*
3.3.2.2 Unidirectional SiC/CAS Composite *142*
3.3.2.3 Unidirectional SiC/CAS-II Composite *144*
3.3.2.4 Cross-Ply C/SiC Composite *144*
3.3.2.5 2.5D C/SiC Composite *149*
3.4 Conclusion *155*
References *156*

4 Interface Damage Law of Ceramic-Matrix Composites *161*
4.1 Introduction *161*
4.2 Interface Damage Law at Room Temperature *163*
4.3 Interface Damage Law at Elevated Temperature in Inert Atmosphere *167*
4.4 Interface Damage Law at Elevated Temperature in Air Atmosphere *172*
4.4.1 1000 °C *172*
4.4.2 1200 °C *175*
4.4.3 1300 °C *178*
4.5 Interface Damage Law at Elevated Temperature in Steam Atmosphere *185*
4.5.1 1000 °C *185*
4.5.2 1200 °C *187*
4.6 Results and Discussion *192*
4.6.1 Effect of Temperature, Oxidation, and Fiber Preforms on Interface Damage of CMCs *192*
4.6.2 Comparisons of Interface Damage Between C/SiC and SiC/SiC Composites *193*
4.7 Conclusion *195*
References *196*

Index *199*

Preface

To realize the advantages of operating systems under high temperature conditions, it is necessary to master the properties of a large number of high temperature materials and components. For example, a significant increase in the gas temperature will significantly increase the gas turbine efficiency. The introduction of new materials and new technology has gradually improved the high-temperature performance of gas turbine engine for more than 70 years, but the development of cooling methods and solutions has contributed more than 75% to the performance improvement. Although component cooling methods and engine material properties have improved significantly, most high-temperature alloys currently operate at temperatures above 90% of their original melting point. Higher operating temperatures are required for more efficient engines, which will require higher component temperatures. As the operating temperature continues to increase, new materials with higher thermo-mechanical and thermo-chemical properties are required to meet high-temperature structural applications. Ceramic-matrix composites (CMCs) are considered to have the potential to provide high strength, high toughness, creep resistance, low notch sensitivity, and environmental stability to meet the needs of future high-performance turbine engines.

For a monolithic ceramic material, when it is subjected to tensile stress, it appears as elastic deformation at low stress level; as the stress increases, cracks occur in the defect region of the material, and the cracks rapidly expand, causing the material to undergo brittle fracture. When the CMC material is subjected to tensile stress, it is elastically deformed before matrix cracks; as the tensile stress increases, the matrix begins to crack, and the fibers begin to debond and play a role of crack bridging; as the tensile stress increases further, the cracks become saturated, and the bridging fibers begin to pull out; as the tensile stress continues to increase, the fibers begin to break until the material reaches the highest strength. The fracture modes of monolithic ceramics and CMCs are different, mainly because the interface plays a role in the fracture process of CMCs. The interface is a special domain between the matrix and the reinforcement. It is the link between the fiber and the matrix, and also a bridge for load transfer. The structure and properties of interphase directly affect the strength and toughness of CMCs. This book focuses on the time-dependent mechanical behavior of CMCs at elevated temperatures, as the following:

(1) The definition, function, and design of interface in different fiber-reinforced CMCs are given. The interphase plays an important role in the mechanical behavior of non-oxide and oxide/oxide CMCs at room and elevated temperatures. The interface phase has two basic functions. One is mechanical fuse function that is to deflect crack growth to protect the fiber, which is the most basic function of the interphase. The second is the load transfer function, which transfers the load to the fiber through shear. In addition to the aforementioned two basic functions, the interphase also plays a buffer role, which is to absorb the residual thermal stress generated due to the mismatch of the thermal expansion coefficient of the fiber and the matrix. The characteristics of pyrolytic carbon (PyC) interphase and boron nitride (BN) interphase used in CMCs are also analyzed.

(2) The effect of the fiber/matrix interface properties and pre-exposure on the tensile and fatigue behavior of fiber-reinforced CMCs is investigated. The experimental tensile and fatigue damage of different CMCs are predicted for different interface properties. The fatigue life S–N curves and fatigue limit stress of unidirectional C/SiC, SiC/CAS (silicon carbide/calcium aluminosilicate) and SiC/1723 (silicon carbide/alkaline-earth aluminosilicate), cross-ply C/SiC, SiC/CAS and SiC/1723, 2D C/SiC and SiC/SiC, 2.5D C/SiC, and 3D C/SiC composites are predicted.

(3) The relationships between the hysteresis dissipated energy and temperature rising of the external surface in fiber-reinforced CMCs under cyclic loading are analyzed. Based on the fatigue hysteresis theories considering fiber failure, the hysteresis dissipated energy and a hysteresis dissipated energy-based damage parameter changing with the increase of cycle number are investigated. The experimental temperature rise-based damage parameter of unidirectional SiC/CAS-II, cross-ply SiC/CAS, and 2D C/SiC composites corresponding to different fatigue peak stresses and cycle numbers are predicted. The fatigue hysteresis behavior of unidirectional, cross-ply, and 2.5D C/SiC composites at room temperature and 800 °C in air atmosphere are investigated.

(4) Comparing experimental fatigue hysteresis dissipated energy with theoretical computational values, the interface shear stress of unidirectional, cross-ply, 2D, and 3D CMCs at room temperature, 600, 800, 1000, 1200, and 1300 °C in inert, air, and steam conditions, are obtained. The effects of test temperature, oxidation, and fiber preforms on the degradation rate of interface shear stress are investigated, and the comparisons of interface degradation between C/SiC and SiC/SiC composites are analyzed.

I hope this book can help the material scientists and engineering designers to understand and master the interface of CMCs.

26 January 2020 *Longbiao Li*

Acknowledgments

I am grateful to my wife Peng Li and my son Li Shengning for their encouragement.

A special thanks to Qian Shaoyu, Pinky Sathishkumar, and Nussbeck Claudia for their help with my original manuscript.

I am also grateful to the team at Wiley for their professional assistance.

1

Definition, Function, and Design of Interface in Ceramic-Matrix Composites

1.1 Introduction

To realize the advantages of operating systems under high-temperature conditions, it is necessary to master the properties of a large number of high-temperature materials and components. For example, a significant increase in the gas temperature will significantly increase the gas turbine efficiency. The introduction of new materials and new technology has gradually improved the high-temperature performance of gas turbine engine for more than 70 years, but the development of cooling methods and solutions has contributed more than 75% to the performance improvement (Li 2018).

Although component cooling methods and engine material properties have improved significantly, most high-temperature alloys currently operate at temperatures above 90% of their original melting point. Higher operating temperatures are required for more efficient engines, which will require higher component temperatures. As the operating temperature continues to increase, new materials with higher thermo-mechanical and thermo-chemical properties are required to meet high-temperature structural applications. Ceramic-matrix composites (CMCs) are considered to have the potential to provide high strength, high toughness, creep resistance, low notch sensitivity, and environmental stability to meet the needs of future high-performance turbine engines (Li 2019).

Figure 1.1 shows the tensile stress–strain curve of monolithic ceramic and fiber-reinforced CMCs. For a monolithic ceramic material, when it is subjected to tensile stress, it appears as elastic deformation at low stress level; as the stress increases, cracks occur in the defect region of the material, and the cracks rapidly expand, causing the material to undergo brittle fracture. When the CMC material is subjected to tensile stress, it is elastically deformed before matrix cracks; as the tensile stress increases, the matrix begins to crack, and the fibers begin to debond and play a role of crack bridging; as the tensile stress increases further, the cracks become saturated, and the bridging fibers begin to pull out; as the tensile stress continues to increase, the fibers begin to break until the material reaches the highest strength. The fracture modes of monolithic ceramics and CMCs are different, mainly because the interface plays a role in the fracture process of CMCs. The interface is a special domain between the matrix

Interfaces of Ceramic-Matrix Composites: Design, Characterization and Damage Effects,
First Edition. Longbiao Li. © 2020 WILEY-VCH GmbH.

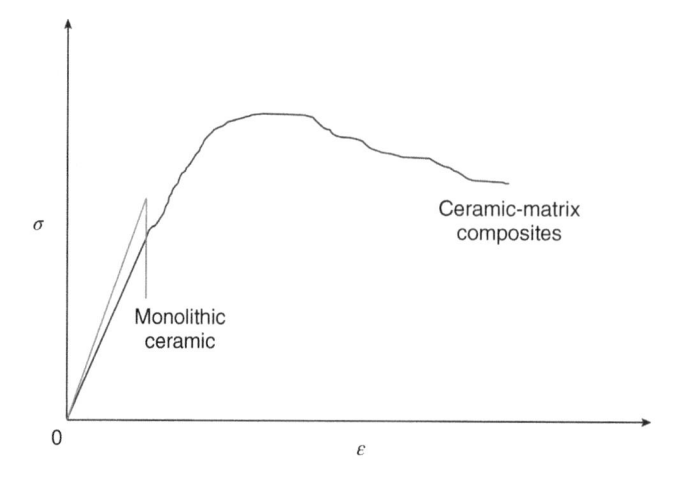

Figure 1.1 The tensile stress–strain curves of monolithic ceramics and fiber-reinforced CMCs.

and the reinforcement. It is the link between the fiber and the matrix, and also a bridge for load transfer. The structure and properties of interphase directly affect the strength and toughness of CMCs.

1.2 The Definition of Interface in Ceramic-Matrix Composites

CMCs possess good damage tolerance, mainly due to frictional sliding at the fiber–matrix interface, and the interphase enables frictional sliding to occur between the fiber and the matrix. The damage tolerance is manifested in CMC with 1% ductility, and the sensitivity to notches is comparable with that of aluminum alloys. CMCs also exhibit good room-temperature fatigue resistance. The fatigue strength (i.e. fatigue failure does not occur when the stress is less than the fatigue strength) is about 90% of the ultimate tensile strength, but the fatigue damage increases with temperature. The tensile strength of CMCs is usually volume-stable as the damage tolerance ability of CMCs decreases the scaling effect that often occurs in the monolithic ceramic materials due to the weakest link. Due to crack deflection and crack tip resistance, CMCs can have cracks that can cause catastrophic failure of monolithic ceramics without fracture. This thermomechanical performance is particularly suitable for producing large, static load, and high-temperature structural components.

CMCs can be divided into two types, i.e. oxide materials and non-oxide materials. Oxide composites include oxide fibers (e.g. Al_2O_3), interphase, and matrix. If any of the aforementioned three components contains non-oxide material (for example, SiC), the composite is classified as non-oxide CMC.

A lot of research and development work is conducted on non-oxide CMCs, especially SiC fiber-reinforced SiC matrix composite (SiC/SiC CMC) using pyrolytic carbon (PyC) or boron nitride (BN) as the fiber interface layer.

Non-oxide CMCs have good high-temperature properties, such as creep resistance and microstructure stability. They also have high thermal conductivity and low thermal expansion, which show a good resistance to thermal stress. SiC/SiC composites are very suitable as thermal load components, such as chamber throat, flap, blade, and heat exchanger. Oxide CMCs (for example, oxide fiber-reinforced porous oxide-matrix composite without interphase) have excellent oxidation resistance, resistance to alkaline corrosion, low dielectric constant, and low price.

Both oxide and non-oxide CMCs exhibit some disadvantages. Non-oxide CMCs (for example, SiC/SiC) exhibit brittleness at intermediate temperatures (about 700 °C). Brittleness is more severe under cyclic loading conditions as oxygen enters from the cracks in the matrix and reacts with the interphase and fibers forming oxidation products, leading to the propagation of the matrix cracks. These oxidation reaction products limit the friction sliding mechanism between the fiber and the matrix inside the material, which can improve the toughness of CMCs. Although this oxidation effect does not occur when the stress is below the proportional limit stress, design and operation experience has shown that it is necessary to consider in advance that the overload stress exceeds the proportional limit stress. Therefore, the local brittleness has become the main limitation of non-oxide CMCs.

Compared with SiC/SiC, oxide CMCs do not undergo oxidation embrittlement, but have a temperature limit (about 1000 °C), which is related to creep and sintering. Moreover, the interface technology of oxide CMCs is less mature than that of non-oxide CMCs. The performance data of most oxide CMCs are obtained in systems without interphase, and the damage tolerance is based on a porous matrix.

1.2.1 Non-oxide CMCs

SiC/SiC CMCs maintain the advantages of SiC matrix such as high-temperature resistance, high strength, low density, and oxidation resistance, realize the strengthening and toughening effect of SiC fiber, and effectively overcome the fatal disadvantages of monolithic ceramics that are brittle, crack sensitivity, and low reliability. Compared with superalloys, SiC/SiC CMCs have lower density (usually 2.0–3.0 g/cm^3, only 1/3–1/4 of superalloys) and higher temperature resistance (>1200 °C under non-cooling conditions) (Liu et al. 2018).

After decades of research, a variety of CMC preparation processes have been developed. Representative processes include chemical vapor infiltration (CVI), polymer infiltration and pyrolysis (PIP) process, and melt infiltration (MI). The main difference between the three processes is the densification of the SiC matrix.

(1) CVI process uses a gaseous precursor (for example, trichloromethylsilane) to pyrolysis and deposit on the surface of the SiC fiber to obtain an SiC matrix.
(2) PIP process usually infiltrates the fiber preform in liquid precursors (such as polycarbosilane), the precursors are ceramicized by high-temperature pyrolysis, and the infiltration and pyrolysis process is repeated to obtain a dense SiC matrix.

(3) Reactive melt infiltration (RMI) is the infiltration of molten silicon into a porous carbon preform, and the reaction of carbon and silicon generates an SiC matrix.
(4) Non-reactive melt infiltration infiltrated the molten silicon into the pores of the matrix, which mainly plays a filling role, and no reaction between silicon and carbon occurs.

The fundamental physical and mechanical properties of SiC/SiC composites prepared by different processes are shown in Table 1.1.

In-plane tensile performance is one of the most important mechanical properties and reflects the strength of the composite material to resist the damage of external tensile load.

Figure 1.2 shows the tensile stress–strain curve of Prepreg-MI SiC/SiC composites (Corman et al. 2016). The curve can be divided into four stages:

- Stage I is the elastic region, and the strain increases proportionally with the stress.
- Stage II is the damage region with a large number of microcracks generated in the matrix of the composite material, and the fibers are debonded. The behavior of matrix cracking and fiber debonding attributes "pseudoplastic" and high toughness of CMCs.
- Stage III is the damage region with saturation of matrix cracking, and the fiber is completely debonded from the matrix.
- Stage IV is where the fiber breaks under higher stress.

The stress–strain curve can be used to obtain Young's modulus, proportional limit stress, ultimate strength, fracture strain, and other related data of the composite material. The proportional limit stress is particularly important, which reflects the maximum stress experienced by the matrix before the generation of matrix microcracks, and is usually defined as the maximum design stress of the component.

There are many factors affected the in-plane tensile performance of SiC/SiC composite, i.e. the type of SiC fiber and fiber volume fraction, as shown in Table 1.2. For the Hi-Nicalon™ S and Tyranno™ SA3 reinforced SiC matrix composites, at the same fiber volume fraction of $V_f = 34.8\%$, the ultimate tensile strength and fracture strain are much higher, and the initial modulus and proportional limit stress are much lower for Hi-Nicalon S SiC/SiC composite than those of Tyranno SA3 SiC/SiC composite. For the SiC/SiC composite with the same fiber and preparation process, i.e. Hi-Nicalon S SiC-SiC MI-CMC or Sylramic™-iBN SiC/SiC PIP-CMC, the ultimate tensile strength increases with fiber volume fraction. The in-plane mechanical properties of SiC/SiC composite are also affected by the temperature. For the Prepreg-MI Hi-Nicalon SiC/SiC composite, when the testing temperature increases from 25 to 1200 °C, the composite initial modulus (E_c) decreases, i.e. from 285 to 243 GPa; the proportional limit stress (σ_{pls}) remains stable, i.e. between 165 and 167 MPa; and the ultimate tensile strength (σ_{UTS}) decreases with increasing temperature, i.e. from 321 to 224 MPa; and the fracture strain (ε_f) is very sensitive to the temperature, and decreases from 0.89% to 0.31%.

Table 1.1 Fundamental physical and mechanical properties of SiC/SiC composites prepared by different processes.

Parameter	CVI	MI					CVI + PIP
	SNECMA	NASA		GE			NASA
	—	N22	N24-A	HiPerComp/Prepreg		HiPerComp/Slurry cast	N26-A
Fiber type	Nicalon	Sylramic	Sylramic-iBN	Hi-Nicalon		Hi-Nicalon	Sylramic-iBN
Fiber volume fraction (%)	40	36	36	22–40		35–38	36
Testing temperature (°C)	23 1000	20	20	25 1200		25 1200	20
Density (g/cm^3)	2.5 2.5	2.85	2.85	2.80 2.76		2.70 2.66	2.52
Porosity (%)	10 10	2	2	<2 —		6 —	14
Thermal conductivity (∥) (W/(m K))	19 15.2	—	—	33.8 14.7		30.8 14.8	—
Thermal conductivity (⊥) (W/(m K))	9.5 5.7	24 (204 °C)	30 (204 °C)	24.7 11.7		22.5 11.8	26 (204 °C)
Coefficient of thermal expansion (∥) (10^{-6}/K)	3 3	15 (1204 °C)	14 (1204 °C)	3.73 (25–1200 °C)		4.34 (25–1200 °C)	10 (1204 °C)
Coefficient of thermal expansion (⊥) (10^{-6}/K)	1.7 3.4	—	—	4.15 (25–1200 °C)		3.12 (25–1200 °C)	—
Initial modulus (GPa)	230 200	250	250	285 243		196 144	200
Proportional limit stress (MPa)	— —	180	180	167 165		120 130	130
Ultimate strength (MPa)	200 200	400	450	321 224		358 271	330
Strain to failure (%)	0.3 0.4	0.35	0.5	0.89 0.31		0.74 0.52	0.40
Interlaminar tensile strength (MPa)	—	—	—	39.5 —		— —	—
Interlaminar shear strength (MPa)	40 35	—	—	135 124		— —	—
Flexural strength (MPa)	300 400	—	—	— —		— —	—
In-plane compressive strength (MPa)	580 480	—	—	1190 >700		— —	—

NASA, National Aeronautics and Space Administration; GE, General Electric Company.

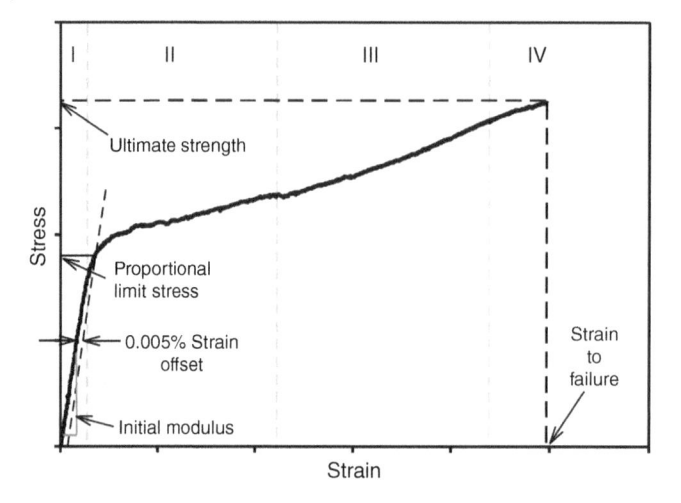

Figure 1.2 Typical tensile stress–strain behavior of Prepreg-MI SiC/SiC composite.

Table 1.2 In-plane tensile properties of SiC/SiC composite with BN interphase at room temperature.

Process	Fiber type	Volume fraction (%)	Elastic modulus (GPa)	Proportional limit stress (MPa)	Ultimate strength (MPa)	Strain to failure (%)
CVI	Hi-Nicalon S	27.7	273	108	273	0.39
	Sylramic-iBN	36.4	242 ± 18	150 ± 5	430 ± 6	0.52 ± 0.017
MI	Hi-Nicalon S	30.2	262	154	341	0.63
	Hi-Nicalon S	34.8	232	147	412	0.60
	Sylramic-iBN	38.9	260 ± 15	184 ± 19	468 ± 30	0.48 ± 0.03
	Tyranno SA3	34.8	254	152	358	0.33
PIP	Sylramic-iBN	52.6	161	133	431	0.35
	Sylramic-iBN	50.0	164	148	317	0.274

The fiber preform also affects the in-plane tensile performance, as shown in Table 1.3. For the 3D orthogonal unbalanced SiC/SiC composite, the fiber volume fraction is the highest along the loading direction (i.e. $V_{fl} = 28\%$), and the ultimate tensile strength is also the highest (i.e. $\sigma_{UTS} > 575$ MPa). For the 3D layer-to-layer angle interlock SiC/SiC composite, the fiber volume fraction is the lowest along the loading direction (i.e. $V_{fl} = 10\%$), and the ultimate tensile strength is also the lowest (i.e. $\sigma_{UTS} = 204$ MPa). When the fiber volume fraction along the loading direction increases, the SiC fiber with high modulus can carry more applied stress, which decreases the stress carried by the matrix and increases the matrix cracking stress. For the 2D triaxial braid composite, the fiber volume fraction along the loading direction is $V_{fl} = 26\%$, which is only slightly lower than that of

Table 1.3 In-plane tensile properties of SiC/SiC composites with different fiber preforms.

Architecture type	Architecture description	Volume fraction in load direction (%)	Elastic modulus (GPa)	Proportional limit stress (MPa)	Ultimate strength (MPa)	Average stress on fibers at failure (MPa)
AI UNI	1D angle interlock through the thickness, Y aligned, unbalanced	23	305 ± 4	322 ± 7	>472	>2052
BRAID	2D triaxial braid (double tow)	26	250 ± 10	245 ± 14	352 ± 18	1352
2D5HS	2D five-harness stain, balanced 0/90 lay-up	19	250	175	463	2362
2D 5HS DT	2D five-harness satin (double tow), balanced 0/90 lay-up	19	197	142	480	2526
LTLAI	3D layer-to-layer angle interlock	10	125	89	204	2040
3DO-Un-R	3D orthogonal, unbalanced	28	275 ± 9	261 ± 16	>575	>2053

3D orthogonal unbalanced composite (i.e. $V_{fl} = 28\%$). However, the tensile ultimate strength is much lower than expected due to the characteristics of shear failure. For the 2D five-harness stain composite, the average stress on fibers at failure is the highest (i.e. 2526 MPa), which indicates that this fiber preform can better realize the loading carry capacity of the fiber.

At high temperature ($T > 900\,°C$), SiC may undergo either active or passive oxidation (Roy et al. 2014; Nasiri et al. 2016). At low oxygen pressure (<1 atm), active oxidation occurs due to the formation of volatile products, as follows:

$$SiC(s) + 2SiO_2(s) \rightarrow 3SiO(g) + CO(g) \qquad (1.1)$$

$$SiC(s) + O_2(g) \rightarrow SiO(g) + CO(g) \qquad (1.2)$$

At high oxygen pressure, passive oxidation occurs, and a protective film of SiO_2 is formed on the surface according to:

$$SiC(s) + \frac{3}{2}O_2(g) \rightarrow SiO_2(g) + CO(g) \qquad (1.3)$$

$$SiC(s) + 2O_2(g) \rightarrow SiO_2(s) + CO_2(g) \qquad (1.4)$$

Oxidation can lead to the degradation of CMC performance, and the degradation behavior is much different for different SiC/SiC composites. For the Prepreg-MI Hi-Nicalon SiC/SiC composite, after exposure at 1200 °C in air atmosphere for 4000 hours, the ultimate tensile strength degrades about 10%. However, after exposure 1000 hours at 1315 °C in air atmosphere, the ultimate tensile strength degrades about 30%. For the Prepreg-MI Hi-Nicalon S SiC/SiC composite, after exposure 4000 hours at 1315 °C in air atmosphere, the ultimate tensile strength degrades less than 10% (Corman et al. 2016).

Ünal et al. (1995) performed oxidation heat-treatment of the composite bars in flowing dried oxygen at 1400 °C for 50 hours. The cut edges of the bars were not sealed after machining, and they were directly exposed to the oxidizing environment, which represents an application case where the matrix is significantly cracked or machined surfaces are exposed. Mechanical testing was conducted at room temperature using a four-point bend fixture with 10 and 20 mm top and bottom span distances, respectively. Oxidation heat treatment of SiC/SiC composites at 1400 °C for 50 hours leads to the formation of the passive oxidation product of cristobalite (SiO_2). However, cristobalite does not seal surface completely, including pores, and there, it is not protective. Mechanical test results clearly demonstrate that oxidation reduces both the fracture stress and the cyclic life, at a given stress level, by about 50%. The degradation of the mechanical properties appears to be related to the preferential oxidation of PyC present at the fiber/matrix interface.

Morscher et al. (2000) investigated the tensile stress-rupture behavior of Hi-Nicalon reinforced MI SiC matrix composites with BN interphase in air at intermediate temperatures. The rupture properties of Hi-Nicalon, BN-interphase, and SiC-matrix composites showed significant loss in load-carrying ability compared with the expected load-carrying ability of the reinforcing fibers for the same temperatures and rupture times. Oxidation of the BN interphase and the formation of borosilicate oxidation products cause strength bonding between individual fibers at locations of near fiber-to-fiber contact. BN interphase oxidation appeared to be enhanced by the carbon layer formed at the Hi-Nicalon fiber surface during matrix processing. At higher stresses, a faster rate for rupture was observed. For this case, through-thickness cracks existed in the matrix due to the initial loading condition. Oxidation ingress occurred from around all sides and edges of matrix cracks into the interior of the composite. Embrittled fibers would fail and shed load onto the remaining fibers. When the load applied to the remaining, unoxidized fibers reached a failure criterion load, based on the reduced load-bearing area, the composite failed. At lower stresses, a slower rupture rate was observed. For this case, microcracks existed in the matrix due to the lower stress loading condition. These cracks would be oxidized and the fibers embrittled. As the embrittled fibers failed due to degradation and strong bonding, the microcracks would grow, shedding the loads of the embrittled fibers onto pristine regions of composite.

Rouby and Reynaud (1993) investigated the tension–tension cyclic fatigue behavior of 2D CVI-SiC/SiC composite at room temperature. The loading frequency is $f = 1$ Hz, and the stress ratio is $R = 0$. The fatigue peak stress is $\sigma_{max} = 50$–175 MPa, and the maximum cycle number is defined as $N = 1\,000\,000$. At room temperature, the fatigue limit stress is $\sigma_{lim} = 135$ MPa, which is 75% of the ultimate tensile strength. According to the different fatigue peak stresses at room temperature, the fatigue damage can be divided into three cases:

- Case I, when the fatigue peak stress is higher than the ultimate tensile strength, the composite is fractured upon first loading to the peak stress.

- Case II, when the fatigue peak stress is between the fatigue limit stress (75% of the ultimate tensile strength) and ultimate tensile strength, fatigue failure occurs within a certain number of applied cycles.
- Case III, when the fatigue peak stress is below the fatigue limit stress, the composite cannot fail within 1 000 000 applied cycles. However, when the fatigue peak stress is higher than the first matrix cracking stress, the shape of the stress–strain hysteresis loops changes with applied cycles.

Chawla et al. (1996) investigated the effect of interphase thickness on tension–tension cyclic fatigue behavior of 2D CVI Nicalon™ SiC/C/SiC composite at room temperature for different loading frequencies. The fibers were woven into a plain-weave fabric and coated with chemical vapor deposited (CVD), PyC. The thickness of the PyC interphase is 0.33 and 1.1 μm. The composites were tested at loading frequencies of $f = 100$ and 350 Hz, and the fatigue peak stress is $\sigma_{max} = 120$ and 150 MPa. Under the same fatigue peak stress, the fatigue life with thickness interphase is longer than that with thin interphase, and the temperature rising of thin interphase is much higher than that of thickness interphase. The composite with thickness interphase protects the fiber from interface wear under high loading frequency, leading to the decrease of surface temperature of composite and increase of fatigue lifetime.

Reynaud et al. (1994) and Reynaud (1996) investigated the cyclic tension–tension fatigue behavior of 2D woven SiC/SiC and cross-ply SiC/MAS-L (silicon carbide/magnesium aluminosilicate-L) composites at elevated temperature in inert atmosphere. For the SiC/SiC composite, the radial stress at the fiber/matrix interface is compressive stress; when the temperature increases, the radial compressive stress decreases, leading to the decrease of the interface shear stress, and the increase of hysteresis dissipated energy. For the cross-ply SiC/MAS-L composite, the radial stress at the fiber/matrix interface is tensile stress, and when the temperature increases, the radial tensile stress decreases, leading to the increase of the interface shear stress and the increase of hysteresis dissipated energy. At elevated temperature between 800 and 1000 °C in inert atmosphere, the chemical reaction occurs at the fiber/matrix interface, leading to the decrease of the interface shear stress. After heat treatment at elevated temperature in inert atmosphere for 50 hours, the interface shear stress decreases through the hysteresis analysis at room temperature.

Elahi et al. (1996) investigated the tension–tension fatigue behavior of 2D CVI Nicalon SiC/SiC composite at room temperature and 1000 °C in air environment. The fatigue life at elevated temperature is much less than that at room temperature. At room temperature, when the fatigue peak stress is 75% of tensile strength, the modulus decreases 50% at initial stage of cyclic loading, then remains stable, and the composite experienced 1 540 000 cycles without fatigue fracture, and the tensile strength remained unchanged, however, the modulus decreased 42%. At elevated temperature, the fatigue peak stress is 77.4% of tensile strength, and the modulus decreased 40% during initial cyclic loading, and then degraded slowly with applied cycles, and when the modulus degradation approached 50%, the composite fatigue fractured.

Ünal (1996a,1996b,1996c) investigated the tensile and fatigue behavior of 2D CVI Nicalon SiC/SiC composite at room temperature and 1300 °C in N_2 atmosphere. The loading frequency is $f = 0.5$ Hz and the stress ratio is $R = 0.1$. The maximum cycle number is defined as 10 000 cycles at room temperature, and 72 000 cycles at 1300 °C. At room temperature, the proportional limit stress is $\sigma_{pls} = 70$ MPa, and the tensile strength is $\sigma_{UTS} = 185$ MPa; however, at 1300 °C, the proportional limit stress is $\sigma_{pls} = 60$ MPa, and the tensile strength is $\sigma_{UTS} = 225$ MPa, and the fracture strain at 1300 °C is twice of that at room temperature, due to the creep damage of fiber. At room temperature, when the fatigue peak stress is $\sigma_{max} = 105$ and 150 MPa, the composite cycled 10 000 without fatigue failure; and when the fatigue peak stress is $\sigma_{max} = 190$ MPa, the composite cycled 422 and fatigue fractured, and the composite modulus degraded rapidly during initial 100 cycles. At elevated temperature of 1300 °C, when the fatigue peak stress is $\sigma_{max} = 72$ and 108 MPa, the composite cycled to the number of $N = 72 000$ without failure; and when the fatigue peak stress is $\sigma_{max} = 145$ and 188 MPa, the composite cycled to 7100 and 890, respectively. The creep damage affected the fatigue failure of SiC/SiC composite at elevated temperature.

Ünal et al. (1995) investigated the oxidation on flexural fatigue behavior of 2D CVI Nicalon SiC/SiC composite. After 50 hours oxidation at 1400 °C, the flexural strength at room temperature decreases from 370 to 189 MPa, and the fatigue limit stress decreases from 283 to 130 MPa, due to the formation of SiO_2 strong interface bonding between the fiber and the matrix.

Mizuno et al. (1996) investigated the tensile and fatigue behavior of 2D CVI Nicalon SiC/SiC composite with 0.1 mm PyC interphase at room temperature and 1000 °C in Ar atmosphere. At room temperature, the proportional limit stress is $\sigma_{pls} = 80$ MPa, and the tensile strength is $\sigma_{UTS} = 251$ MPa; and at elevated temperature in Ar atmosphere, the proportional limit stress is $\sigma_{pls} = 100$ MPa, and the tensile strength is $\sigma_{UTS} = 260$ MPa. The tensile strength and failure strain at elevated temperature are higher than those at room temperature, due to the degradation of interface shear stress at elevated temperature in inert atmosphere. However, at elevated temperature in air atmosphere, the tensile strength and fracture strain are both lower than those at room temperature, and the pullout length at the fracture surface is also shorter than that at room temperature, due to the formation of strong interface bonding after interface oxidation. For the cyclic fatigue testing, the loading frequency is $f = 10$ Hz at room temperature, and $f = 20$ Hz at elevated temperature, and the stress ratio is $R = 0.1$, and the maximum cycle number is defined as 10 000 000. At room temperature, the fatigue limit stress is $\sigma_{fls} = 160$ MPa, which is higher than the proportional limit stress (i.e. $\sigma_{pls} = 80$ MPa); and at elevated temperature, the fatigue limit stress is $\sigma_{fls} = 75$ MPa, which is much lower than the proportional limit stress ($\sigma_{pls} = 100$ MPa). The degradation of fatigue limit stress at elevated temperature is attributed to the fiber creep and interface wear.

Zhu et al. (1996, 1997, 1998, 1999) and Zhu and Kagawa (2001) investigated the tensile and fatigue behavior of 2D CVI SiC/SiC composite at room temperature and 1000 °C in Ar atmosphere. At room temperature, the fatigue limit stress is 70–80% tensile strength, which is much higher than the first matrix cracking stress. Under cyclic fatigue loading, the propagation of matrix cracking is affected by the bridging fibers, and the stress intensity factor decreases due to the bridging effect of the fiber. However, at 1000 °C in Ar atmosphere, the fatigue life decreases greatly, and the fatigue limit stress is $\sigma_{pls} = 75$ MPa, which is about 30% of the tensile strength at elevated temperature.

Zhu (2006) investigated the effect of oxidation of fatigue behavior of 2D CVI SiC/SiC composite. After oxidation for 100 hours at 600 °C, the PyC interphase between the fiber and the matrix disappears, leading to 13% degradation of fatigue life; and after oxidation for 100 hours at 800 °C, the strong interphase of SiO_2 appears between the fiber and the matrix, leading to the shorter fatigue life at room temperature.

Zhu et al. (2004) investigated the low-cyclic fatigue behavior of 3D Tyranno SiC/[Si–Ti–C–O] composite at room temperature. The loading frequencies were $f = 0.02$, 0.2, and 20 Hz, and the fatigue stress ratio was $R = 0.1$ and -0.1. At low loading frequency, the stress corrosion is the main reason for the low-cyclic fatigue failure; and at high loading frequency, the PyC interphase is easy to wear, and the interface debonding length increases with cycles for interface wear, which increases the load carried by the fibers and also decreases the fiber strength, leading to the final fatigue fracture.

Kaneko et al. (2001) and Zhu et al. (2002) investigated the effect of loading frequency on fatigue life of 3D CVI and PIP Tyranno SiC/SiC composites at room temperature. The fibers were coated with nano-scale carbon to decrease interface bonding between fiber and matrix. This coating acts to increase strength and toughness of the composites. The fatigue tests were carried out under load control with a sinusoidal loading frequency of $f = 20$ and 0.2 Hz (PIP SiC/SiC composite) and $f = 20$, 0.2, and 0.02 Hz (CVI SiC/SiC composite) and a stress ratio of $R = 0.1$. There is no difference in fatigue life between 20 and 0.2 Hz for the CVI SiC/SiC composite, but the fatigue life at 0.2 Hz is slightly lower than that at 20 Hz for the PIP SiC/SiC composite, and the fatigue life at 0.02 Hz is much lower than those at 20 and 0.2 Hz for CVI SiC/SiC composite. The fatigue life for PIP SiC/SiC composite is time-dependent at 20–0.2 Hz. For CVI SiC/SiC composite, the fatigue life is time-dependent at 0.2–0.02 Hz, but cyclic-dependent at 20–0.2 Hz. Modulus reduction of PIP SiC/SiC composite saturates after 10 cycles, however, for CVI SiC/SiC composite, the modulus gradual decreases with number of cycles. Because of low sliding stress of PIP SiC/SiC composite, the fatigue strength of PIP SiC/SiC composite is much higher than that of CVI SiC/SiC composite.

Ruggles-Wrenn et al. (2011, 2018) investigated the effect of loading frequency and environment on the fatigue performance of 2D CVI Hi-Nicalon SiC/SiC, CVI Hi-Nicalon SiC/[SiC-B_4C], and MI Hi-Nicalon SiC/SiC composites. The CVI Hi-Nicalon SiC/SiC composite was processed by CVI of SiC into the fiber preforms. Before the infiltration, the fiber preforms were coated with BN fiber coating (~0.25 μm thick) to decrease bonding between the fibers and the matrix. The CVI Hi-Nicalon SiC/[SiC-B_4C] composite had an oxidation inhibited matrix consisting of alternating layers of SiC and B_4C. Prior to infiltration, the fiber preforms were coated with PyC fiber coating (~0.4 μm thick) with boron carbide overlay (~1.0 μm thick) to create a weak fiber/matrix interface. The fiber preforms of MI Hi-Nicalon SiC/SiC composite were coated with a CVI BN interphase coating, then a CVI SiC coating of initial matrix was applied to rigidize the preforms and to protect the fibers, followed by slurry infiltration of SiC particulates and infiltration of molted Si to fill in the remaining porosity. Under cyclic fatigue loading at 1200 °C in air or in steam condition with the loading frequency of 0.1, 1.0, and 10 Hz, the fatigue life decreases with increasing loading frequency. The material performance degraded rapidly in steam condition. After cyclic fatigue loading in air condition without fatigue failure, the tensile strength at room temperature remained stable, and Young's modulus decreased up to 22%, which indicated that the fiber remained undamaged under cyclic fatigue loading, and the matrix cracking occurred.

The creep property of CMCs is closely related to their constituents and preparation process. For different fiber-reinforced same matrices, the creep property of composite is consistent with the reinforcing fiber. The higher the creep resistance of the fiber, the better the creep resistance of the composite (Morscher 2010). For Tyranno SA, Hi-Nicalon S, and Sylramic-iBN reinforced MI-SiC matrix, the creep resistance of SiC/SiC composite is consistent with that of fiber, i.e. Tyranno SA < Hi-Nicalon S ≤ Sylramic-iBN (Bunsell and Berger 2000; DiCarlo and Yun 2005). For the same fiber-reinforced different densification process matrix, for example, Sylramic-iBN reinforced MI, CVI, PIP SiC matrix, the creep resistance of PIP SiC/SiC composite is the worst, mainly due to the matrix microcracks during fabrication, which decreases the load carrying capacity of the matrix. For the same densification process matrix reinforced by the same fiber, if the matrix components are different, the creep performance of the composite will also be different. For example, for the CVI-G and CVI-H matrix reinforced by Sylramic-iBN fiber, the creep fracture time of the former is only 25 hours at 138 MPa, and the latter is 250 hours, because the matrix of CVI-G is silicon rich, leading to the more prone to creep of the matrix.

Morscher et al. (2008) investigated the retained properties, damage development, and failure mechanisms of 2D MI Sylramic-iBN/BN/SiC composite under tensile creep and fatigue loading. The results show that the retained room temperature tensile strength and modulus decrease with the increase of stress and time. Under different experimental conditions, the failure mechanisms of degradation are different. Under high stress and short time, the oxidation-induced unbridged crack growth emanating from composite surface causes load redistribution in intact region and local stress concentration at matrix crack tip, and eventually one of these cracks develops into fracture pattern with time. Under low stress and long time, the fiber strength degradation is probably due to an intrinsic creep-controlled flaw growth mechanism or attack of the fibers from Si diffusion through the CVI SiC portion of the matrix. However, the matrix cracking stress increases after creep or fatigue test because of the increase of the residual compressive stress of the matrix after the test. For example, for the original composite, the residual compressive stress of the matrix is about 50 MPa, but after tensile creep, the value increases to more than 100 MPa, which is mainly related to the relaxation of the matrix during creep. After unloading, the fiber generates greater compressive stress on the matrix. Therefore, it is necessary to apply more stress to the formation and propagation of matrix cracks to overcome the existing compressive stress. Although creep or fatigue tests increase the matrix cracking stress, they also increase the number and depth of matrix cracks, which eventually lead to material damage evolution.

In the combustion environment, the fatigue and creep performance and degradation mechanisms are much different from those under air environment. Kim et al. (2010) and Sabelkin (2016) investigated the fatigue and creep performance of 2D MI Hi-Nicalon S SiC/BN/SiC composite under combustion environment and compared with the performance at elevated temperature in air condition. The fatigue life was greater in air condition, approximately by an order of magnitude than in the combustion environment at the same applied peak stress. For example, the fatigue life at the same applied peak stress of 125 MPa was significantly higher in air condition than in the combustion environment, i.e. 58 838 versus 8329 cycles. The creep life decreases about 86% in combustion environment, due to the higher water vapor content in the combustion environment than that in the air condition. The higher water vapor content accelerates the oxidation of BN interphase. At the same time, under the burning of the combustion flame, the compression stress is generated on the front side of the sample, and the tensile stress is generated on the back side. These local thermal stresses reduce the stress threshold of the formation and propagation of the matrix crack, which makes the matrix more prone to crack, providing the channel for the diffusion of oxygen and aggravating the degradation of composite properties.

The degradation behavior of SiC/SiC composites in combustion environment is also affected by the combustion flame temperature and the surface texture of the composite. Bertrand et al. (2015) investigated the tension–tension fatigue behavior of 2D CVI Sylramic-iBN/BN/SiC composite under combustion environment. Fatigue tests were performed at a stress ratio of 0.1 and loading frequency of 1 Hz. The combustion environment was created using a high-velocity oxygen fuel gun, which impinged the flame directly on the one side of specimen when it was subjected to cyclic load. The flame-impinged surface of the specimen was heated top the average temperature of 1250, 1350, and 1480 °C. The SiC/SiC composite achieved a run-out of 25 hours at 46% and 33% of tensile strength under combustion environment at 1250 and 1350 °C, respectively. However, run-out was not achieved at 1480 °C due to erosion and degradation of material. Oxidation embrittlement occurred near the surface on both flame side and backside of the specimen, but more on the flame side. It appears that the make-up of the material, i.e. uneven surface texture, contributed to the increased temperature locally which contributed to the erosion and degradation of the material. The surface texture can be modified with the use of environmental barrier coatings.

1.2.2 Oxide/Oxide CMCs

Oxide/oxide composites possess low density, high-temperature resistance, and oxidation resistance (Zok 2006; Yang et al. 2018). Compared with SiC/SiC composites, oxide/oxide composites have better environmental stability (Lebel et al. 2017; Singh et al. 2017a,2017b), which can serve in the combustion environment

Table 1.4 Properties of oxide fibers.

Fiber	Company	Mass fraction (%) Al$_2$O$_3$	Mass fraction (%) SiO$_2$	Mass fraction (%) Other	Diameter (μm)	Strength (MPa)	Modulus (GPa)	Density (g/cm³)
Nextel 312	3M	63	25	B$_2$O$_3$	10–12	1700	150	2.7
Nextel 440	3M	70	28	B$_2$O$_3$	10–12	2000	190	3.1
Nextel 550	3M	73	27	—	10–12	2000	193	3.0
Nextel 610	3M	> 99	<0.3	Fe$_2$O$_3$	10–12	3100	380	3.9
Nextel 650	3M	89	—	ZrO$_2$, Y$_2$O$_3$	10–12	2550	358	4.1
Nextel 720	3M	85	15	—	10–12	2100	260	3.4
FP	DuPont	99	—	—	15–25	1400–2100	350–390	3.59
PRD-166	DuPont	80	—	ZrO$_2$	15–25	2200–2400	85–120	—
Altex	Sumitomo	85	15	—	10/15	1800	210	3.3
Nitivy ALF	Nitivy	72	28	—	7	2000	170	2.9
Saffil	ICI	95	5	—	3	1030	100–300	2.8–3.3

of 1000–1400 °C for a long time (van Roode and Bhattacharya 2013; Askarinejad et al. 2015; Kiser et al. 2015; Lanser and Ruggles-Wrenn 2016; Behrendt et al. 2016).

The preparation methods of oxide/oxide composites include sol–gel method (Sol–Gel), CVI, RMI, PIP process, gel casting (Gelcasting) process, electrophoretic deposition (EPD), etc.

The reinforcing fibers of oxide/oxide composite include Al$_2$O$_3$ and Al$_2$O$_3$-SiO$_2$ ceramic fibers. Among commercial oxide fibers, Nextel™ series produced by 3M Company of the United States is the most mature and widely used (Wilson and Visser 2001). In addition, there are FP and PRD-166 series of DuPont company, Altex series of Sumitomo company of Japan, Nitivy ALF series of Nitivy company, and Saffil series of ICI company of the United Kingdom. The basic properties of commonly used oxide fibers are shown in Table 1.4 (Krenkel 2008).

Nextel 312 of 3M Company is the first continuous alumina fiber in the world. Its composition contains Al$_2$O$_3$, SiO$_2$, and B$_2$O$_3$. Due to the appearance of glass phase in the fiber, its creep performance is significantly affected, which limits its maximum service temperature. To improve the high-temperature stability of the oxide fiber, 3M Company further reduced the content of B$_2$O$_3$ in Nextel 312, and developed Nextel 440, which can be applied to the condition at temperature below 1000 °C, for example, thermal insulation environment. Nextel 550 fiber contains only γ-Al$_2$O$_3$ and amorphous SiO$_2$, and its service temperature is further improved, but it is generally used in the condition at temperature below 1200 °C due to the crystallization temperature of the fiber. To meet the requirements of high-temperature stability of CMCs for hot section components in aerospace, 3M Company has developed Nextel 610 fiber. Nextel 610 is almost completely composed of α-Al$_2$O$_3$. At room temperature, it has tensile strength up to 3100 MPa. A small amount of SiO$_2$ is added to Nextel 610. At high-temperature, it can react with Al_2O_3 to form mullite and wrap it on the surface of Al_2O_3 grains to prevent the growth of Al_2O_3 grains. Therefore, Nextel 610 still has a strength retention of more than 90% in the high-temperature environment of 1200 °C, but the fiber is easy to creep fracture at temperature above 1300 °C. To improve the creep resistance of the fiber, 3M Company developed Nextel 720™ fiber. Nextel 720 contains 45% α-Al_2O_3 and 55% mullite. Mullite has extremely excellent creep resistance, so the creep resistance of fiber is greatly improved. In addition, 3M Company also developed Nextel 650 fiber, which is mainly composed of α-Al_2O_3. Adding a small amount of ZrO_2 and Y_2O_3 can inhibit the growth of grains and reduce the creep rate. Nextel 650 fiber has better high-temperature tensile strength than Nextel 720, and better creep resistance than Nextel 610.

The matrix materials of oxide/oxide composites mainly include alumina (mainly α-Al_2O_3), mullite ($3Al_2O_3$-$2SiO_2$), yttrium aluminum garnet ($Y_3Al_5O_{12}$ [YAG]), lithium aluminum silicon (LAS), barium aluminum silicon (BAS) glass, etc.

α-Al_2O_3 has moderate sintering temperature, high melting point, excellent mechanical properties, chemical corrosion resistance, and excellent high-temperature oxidation resistance and is widely used as the matrix material of oxide/oxide composites (Ben Ramdane et al. 2017). Ruggles-Wrenn et al. (2006, 2008a,2008b,2008c), Ruggles-Wrenn and Szymczak (2008), Ruggles-Wrenn and Braun (2008), Ruggles-Wrenn and Laffey (2008), Ruggles-Wrenn and Whiting (2011), Lanser and Ruggles-Wrenn (2016), and Ruggles-Wrenn and Lanser (2016) conducted a large number of performance tests on alumina matrix oxide/oxide composites prepared by ATK-COI Ceramic Company, and the results showed that their overall mechanical properties are excellent, but their high-temperature creep resistance is poor. It is easy to cause creep damage of composite materials at elevated temperature.

Mullite ($3Al_2O_3$-$2SiO_2$) is a series of minerals composed of aluminosilicates. It has high melting point, low density, small linear expansion coefficient, stable high-temperature physical and chemical properties, and excellent creep and thermal shock resistance.

The design of oxide/oxide composites mainly uses two basic principles. One is to use an interface layer, generally using a fiber coating; the other is to use a sufficiently weak matrix such as a porous matrix. The fracture behavior of composites is the result of competitive fracture between fibers, interface layers, and the matrix. Tough composites are designed to achieve crack deflection at or near the fiber/matrix interface.

$LaPO_4$ is the most common weak oxide interface layer that meets requirements of crack deflection. $LaPO_4$ has a high melting point (>2000 °C), its bonding with oxides, especially alumina, is weak, and it can coexist stably with oxides such as alumina at high temperatures. Morgan and Marshall (1995) tested the sapphire/$LaPO_4$/alumina composite system and found that the matrix cracks did not penetrate into the fiber, but deflected at the $LaPO_4$/fiber interface. Keller et al. (2003) found that the Nextel 610/Al_2O_3 composite with $LaPO_4$ interface layer has higher strength and service temperature. After heating at 1200 °C for 100 hours, the strength loss of the composite with interface layer is about 28%. After heating at 1200 °C for 1000 hours, the strength retention of the composite still exceeds

60%. However, the strength loss of the composite without interface layer after heating at 1200 °C for five hours is more than 70%. The fracture surface of Nextel 610/LaPO$_4$/Al$_2$O$_3$ composite shows that the fiber is pulled out from the alumina matrix, and LaPO$_4$ exists on the pulled fiber surface. However, the composites without interface layer show brittle fracture, and no fiber is pulled out.

The porous interface layer has a pore structure, and the micropores can effectively deflect matrix cracks, thereby consuming the fracture energy of the composite material. Holmquist et al. (2000) carried out research on sapphire/porous ZrO$_2$/alumina composites. Microcracks existed in the matrix, while the fibers remained intact, which proved that the porous ZrO$_2$ coating is an effective weak interface layer.

Fugitive coating refers to a type of coating that the interface layer can be removed during the preparation of the composite material. The carbon interface layer in the oxide/oxide composite can be removed by oxidation before or during use, leaving gaps at the fiber/matrix interface. Keller et al. (2000) showed that fugitive carbon interface layers can provide weak interface layers for sapphire/YAG and Nextel 720/calcium aluminosilicate (CAS) composites.

In the design of oxide/oxide composites, a relatively weak matrix is currently widely used to replace the fiber/matrix interface layer. In porous matrix composites, cracks in the matrix are deflected in the matrix at the fiber/matrix interface. Compared with a dense matrix, the porous matrix does not cause stress concentration on the fiber surface to break the fiber. Although the bond between the fibers and the matrix particles is strong, cracks generally develop toward adjacent pores and eventually reach the fiber surface in the pores, resulting in deflection.

Steel et al. (2001) investigated the tensile and tension–tension fatigue behavior of PIP 8HSW Nextel 720/alumina composite at room temperature and 1200 °C in air atmosphere. The fatigue tests were under load control with the loading frequency of $f = 1$ Hz, stress ratio of $R = 0.05$, and the maximum applied cycle number of $N = 100\,000$. At room temperature, the tensile strength is $\sigma_{UTS} = 144$ MPa, and the fatigue limit stress is $\sigma_{pls} = 102$ MPa, which is 70% of tensile strength. At 1200 °C in air condition, the tensile strength is $\sigma_{UTS} = 140$ MPa, and the fatigue limit stress is $\sigma_{pls} = 122$ MPa, which is 87% of tensile strength. The fatigue damage mechanisms at elevated temperature are similar with those at room temperature; however, the fiber creep also affects the fatigue damage.

Zawada et al. (2003) investigated the tensile, shear, fatigue and creep behavior of 2D Nextel 610/Al$_2$O$_3$-SiO$_2$ composite at room temperature and 1000 °C in air atmosphere. The fatigue loading frequency is $f = 1$ Hz, stress ratio is $R = 0.05$, and maximum applied cycle is $N = 100\,000$. At room temperature, the fatigue limit stress is $\sigma_{pls} = 170$ MPa, which is 85% of tensile strength; at 1000 °C in air atmosphere, the fatigue limit stress is $\sigma_{pls} = 150$ MPa, which is 85% of tensile strength. At elevated temperature, when the fatigue peak stress is between $\sigma_{max} = 100$ and 150 MPa, the modulus decreases 5–10% during first 1000 cycles, and then the modulus remains constant till the maximum applied cycle number, which indicates that there is no obvious damage accumulation under cyclic fatigue loading. The energy dissipation for each cycle is very low, only 3–5 kJ/m^3. When the fatigue peak stress is $\sigma_{max} = 150$ MPa, the strain range remains unchanged with increasing cycles, which indicates that there is little damage accumulation during cyclic fatigue loading. After experiencing 100 000 cycles under $\sigma_{max} = 150$ MPa at elevated temperature, the tensile strength at room temperature is the same with that of original specimen; however, the modulus decreases mainly due to matrix cracking. The oxide/oxide composite has good fatigue and oxidation resistance at high temperatures.

Mall and Ahn (2008) investigated the effect of loading frequency on tension–tension fatigue behavior of 2D Nextel 720/alumina composite at room temperature. The fatigue tests were controlled under load with the stress ratio of $R = 0.05$, and the loading frequency of $f = 1, 100,$ and 900 Hz, and the corresponding maximum cycle number is $N = 100\,000, 100\,000\,000,$ and $100\,000\,000$. When the loading frequency increases from $f = 1–900$ Hz, the fatigue life increases with loading frequency. When the fatigue peak stress is $\sigma_{max} = 120$ MPa, the fatigue cycle number is $N = 600, 6000,$ and $100\,000\,000$ for the loading frequency of $f = 1, 100,$ and 900 Hz. Fatigue behavior of Nextel 720/alumina appeared to be a combination of cycle-dependent and time-dependent phenomena. Surface temperature of specimens tested at 900 Hz increased considerably relative to that at 1 or 100 Hz (70 °C versus 5 °C increase). Damage mechanisms showed an evidence of local fiber/matrix interfacial bonding developed during cycling due to frictional heating at the highest frequency of 900 Hz. This was not observed at the two lower frequencies. This interfacial bonding may have caused an increase in fatigue life/strength of the tested CMC system at the highest frequency of 900 Hz.

Ruggles-Wrenn et al. (2008a) investigated the effect of loading frequency on fatigue life of 2D Nextel 720/alumina composite at 1200 °C in air and in steam environment. In air condition, the loading frequency is $f = 0.1$ and 1 Hz, and the fatigue peak stress is between $\sigma_{max} = 100$ and 170 MPa, and the maximum applied cycle number is $N = 100\,000$; in steam atmosphere, the loading frequency is $f = 0.1, 1,$ and 10 Hz, and the fatigue peak stress is between $\sigma_{max} = 75$ and 170 MPa, and the maximum applied cycle number for the loading frequency of $f = 0.1$ and 1 Hz is $N = 100\,000$, and the maximum applied cycle number of the loading frequency of $f = 10$ Hz is $N = 1\,000\,000$. In air condition, the loading frequency of $f = 0.1$ and 1.0 Hz has little effect on the fatigue life, and the fatigue limit stress is $\sigma_{pls} = 170$ MPa, which is 88% of the tensile strength. After fatigue loading, the residual strength remains unchanged, however, the modulus decreases about 30%. In steam condition, the loading frequency affects the fatigue life. When the loading frequency is $f = 10$ Hz, the fatigue limit stress is $\sigma_{pls} = 150$ MPa, which is 78% of the tensile strength. After fatigue loading, the residual strength decreases about 4%, and the modulus decreases about 7%. When the loading frequency is $f = 1$ Hz, the fatigue limit stress is $\sigma_{pls} = 125$ MPa, which is 69% of the tensile strength. After fatigue loading, the residual strength decreases about 12%, and the modulus decreases about 20%. However, for the loading frequency of $f = 0.1$ Hz, the fatigue limit stress is lower than 75 MPa. On the fracture surface, there exists a lot of fiber pullout for the loading frequency of $f = 10$ Hz, and few fiber pullout for the loading frequency of $f = 0.1$ Hz.

1.3 The Function of Interface in Ceramic-Matrix Composites

In CMCs, the interphase is usually used to control the bonding strength between the fiber and the matrix. The mechanical behavior of CMCs depends not only on the properties of the fiber and the matrix, but also on the interface bonding strength between the fiber and the matrix. If the bonding strength between the fiber and the matrix is too high, the stress concentration will make the fiber unable to carry load uniformly, resulting in low strength of CMCs. At the same time, due to the lack of energy absorption mechanism in the process of fracture, the toughness of CMC is low. Therefore, only when the bonding strength between the fiber and the matrix is moderate, the fiber can not only effectively bear the load, but also consume energy through debonding and pulling out in the process of fracture, so as to realize the strengthening and toughening of CMC. Because of the existence of the interface layer, the CMCs can have ductile fracture. The toughening mechanisms of CMCs include crack deflection, fiber debonding, fiber bridging, and fiber pullout. Fiber pullout is the main toughening mechanism of composite materials. Through the friction energy consumption in the process of fiber pullout, the fracture work of composite materials increases, while the energy consumption in the process of fiber pullout depends on the fiber pullout length and the sliding resistance of debonding surface. The sliding resistance is too large, the fiber pullout length is short, the toughening effect is not good, and the strength is low. The pullout length of the fiber depends on the strength distribution of the fiber and the sliding resistance of the interface, while the sliding resistance mainly depends on the interface layer and the surface roughness of the fiber.

Under tensile loading of CMCs, matrix cracking and interface debonding occur, and the fiber and matrix stress distribution can be determined using the shear-lag theory.

When the stress in the matrix approaches the matrix strength, matrix will crack. When the interface completely debonds, the stress in the matrix cannot approach the matrix strength, and the matrix cracking approaches saturation. During matrix cracking evolution, the process includes the first matrix cracking and the saturation of matrix cracking.

For the CMC system, the core of its strengthening and toughening is the optimal design of interphase. The interface phase has two basic functions. One is mechanical fuse function that is to deflect crack growth to protect the fiber, which is the most basic function of the interphase. The second is the load transfer function, which transfers the load to the fiber through shear. In addition to the aforementioned two basic functions, the interphase also plays a buffer role, which is to absorb the residual thermal stress generated due to the mismatch of the thermal expansion coefficient of the fiber and the matrix. In order to achieve this function, the interphase must be thick enough to have a compliant. Because most CMCs are used in high-temperature oxidizing atmosphere, the microcracks in the matrix will accelerate the diffusion of oxidizing atmosphere to the internal interphase and fibers, which requires the interphase to have oxidation resistance.

1.3.1 Effect of Interphase on Sliding Resistance

The introduction of interphase can change the roughness of debonding surface, the interphase can change the interaction surface between fiber and matrix, and the increase of interphase thickness can weaken the engagement strength in the process of sliding, and reduce the frictional resistance of sliding. If the interphase thickness exceeds the roughness amplitude, the friction stress can be almost eliminated. In addition, the crystallization of the interphase after high-temperature treatment will increase, for example, the graphitization of the PyC interphase after heat treatment will increase, which will also help the interface slip and reduce the slip resistance.

1.3.2 Effect of Interphase on Thermal Misfit Stress

Interphase can adjust the thermal residual stress (TRS) between the fiber and the matrix. The thermal expansion coefficient and thickness of interphase affect the TRS (Kuntz et al. 1993).

Ignore the influence of the geometry and the elastic constants of the fiber; the fiber radial TRS can be determined using the following equation.

$$\sigma_r = E_m \Delta T \Delta \alpha \tag{1.5}$$

The factors in Eq. (1.5) are defined as follows: $\Delta T = T - T_0$, where T_0 is the stress-free temperature (e.g. the fabrication temperature) and T is the temperature at which the stress is determined. Usually, $\Delta T < 0$, when T_0 is the fabrication temperature. $\Delta \alpha = \alpha_m - \alpha_f$, where α is the thermal expansion coefficient and the subscripts f and m refer to the fiber and the matrix, respectively. The radial stresses become negative (pressure) when $\alpha_m > \alpha_f$ or positive (tension), when $\alpha_m < \alpha_f$. E_m is Young's modulus of the matrix. Faber (1997), Kuntz et al. (1993), Davis et al. (1993), Brennan (1990), and Kerans (1996) developed methods to calculate the TRS; however, these calculations were conducted without considering the thickness of the interphase.

For the SiC/PyC/borosilicate glass composite, the modulus of interphase is about 80 GPa, and the thickness is about 80 nm. The thermal expansion coefficient of PyC interphase has a significant effect on the axial and circumferential residual stresses in PyC, and the effect on the radial residual stresses can be ignored. The thermal expansion coefficient of interphase has a significant effect on the TRS in fiber and interphase, but the effect on matrix TRS is small (Kuntz et al. 1993).

1.4 The Design of Interface in Ceramic-Matrix Composites

Interphase plays an important role in CMCs, but not all high-temperature materials can be used as interphase materials, which need to meet three conditions:

(1) Low modulus. The low modulus can reduce the mismatch of thermal expansion coefficient and modulus between the fiber and the matrix, thus reducing the physical damage of fiber.
(2) Low shear strength. Because the interface is the place where the matrix cracks deflect, the interphase should have low shear strength, which can realize the crack deflection, fiber pullout, debonding, and other toughening mechanisms.
(3) High thermal and chemical stability. It can prevent the fiber from being damaged in the process of composite preparation, and it can exist stably in the process of composite service.

There are few interphase materials that can meet the above conditions. This kind of materials is usually composed of layered crystal materials. The bonding force between layers is weak, and the direction of the lamellae is parallel to the fiber surface. Commonly used are PyC interphase and BN interphase.

1.4.1 PyC Interphase

PyC is a traditional and common interphase of CMCs, and it is the first interphase that can produce toughness in ceramic fiber-reinforced CMCs. The in situ growth of carbon interphase on the fiber surface can make the weak interface between the fiber and the matrix, thus improving the fracture toughness of the composite. The toughness of the composite increases with the increase of the thickness of the interphase, but the load transfer between the matrix and the fiber decreases with the increase of interphase thickness. Therefore, the interphase thickness is generally 0.1–0.3 μm.

The use of a PyC interphase can significantly improve the performance of composite materials at room temperature or in a high-temperature inert atmosphere, but PyC begins to oxidize at 400 °C in air atmosphere, and CMC materials are mostly used in air atmosphere. The PyC interphase is not an ideal interphase. Generally, the PyC interphase requires the matrix to have a high dense level, so that the fiber does not directly contact the outside oxidizing atmosphere, or one or more dense protective layers are deposited on the PyC surface. In the case of cracking occurrence, the PyC interphase is still easy to damage.

Filipuzzi and Naslain (1994) and Filipuzzi et al. (1994) studied the interphase oxidation in Nicalon SiC/PyC/SiC composites. Carbon is oxidized to generate CO and CO_2 gas. After the gas is released, a gap is left between the fiber and the matrix, causing adjacent fibers and the matrix to oxidize and generate SiO_2. Oxygen diffuses along the voids that occur during the oxidation of the fibers and the matrix, resulting in continuous oxidation of the composite material until the SiO_2 generated by the oxidation of the fibers and the matrix closes the voids. According to the oxidation rate constants of Nicalon fibers and SiC matrix, Luthra (1994) calculated the time required to close the voids and found that it increase with the increase in the thickness of the interphase.

Non-oxide CMCs are designed to operate below the proportional limit stress, but it should be considered that they sometimes operate at temperatures above the proportional limit stress. When this happens, oxygen diffuses along the matrix cracks and oxidizes the PyC interphase.

1.4.2 BN Interphase

In addition to PyC, BN is the only fiber coating capable of slow failure in non-oxide composites. It should be noted that not all BNs are the same. When deposited at low temperature, the generated BN is amorphous; when deposited at high temperature (for example, above 1500 °C), BN generally has an ordered hexagonal crystal structure (which can make CMCs exhibit tough behavior). The deposited BN must contain carbon or oxygen; and the silicon-doped BN has higher oxidation resistance.

As CMCs with PyC interphase, the toughness of CMCs with BN interphase increases with coating thickness. The optimal thickness range of BN interphase should be 0.3–0.5 μm, which is thicker than PyC interphase.

Since the oxidation products of BN are liquid boron oxide rather than gaseous oxide, it is expected that the oxidation rate of BN in dry air or oxygen is much lower than that of carbon. The oxidation depth of the BN coating of the SiC-Si matrix composite material observed in the experiment is very shallow. When the oxidation is performed at a temperature range of 700–1200 °C for 100 hours, the oxidation depth is only in the order of 10 μm or less. The oxidation depth of BN coating is much shallower than that of PyC coating. In dry oxidizing environments, BN coatings perform better than PyC coatings.

Oxidation experiments in a humid environment show that the oxidation properties of BN coatings are similar to those of carbon coatings. When the matrix is cracked, oxidants (oxygen and water vapor) can quickly diffuse through the matrix cracks and oxidize the fiber coating. Generally the thickness of the coating does not exceed 1 μm. As a result, the coating will oxidize rapidly, only a few minutes at an experimental temperature above 900 °C (Jacobson et al. 1999), and this damage will quickly spread to the fibers.

1.5 Conclusion

In this chapter, the definition, function, and design of interface in different fiber-reinforced CMCs are given. The interphase plays an important role in the mechanical behavior of non-oxide and oxide/oxide CMCs at room and elevated temperatures. The interface phase has two basic functions: one is mechanical fuse function that is to deflect crack growth to protect the fiber, which is the most basic function of the interphase. The second is the load transfer function, which transfers the load to the fiber through shear. In addition to the aforementioned two basic functions, the interphase also plays a buffer role, which is to absorb the residual thermal stress generated due to the mismatch of the thermal expansion coefficient of the fiber and the matrix. The characteristics of PyC interphase and BN interphase used in CMCs are also analyzed.

References

Askarinejad, S., Rahbar, N., Sabelkin, V., and Mall, S. (2015). Mechanical behavior of a notched oxide/oxide ceramic matrix composite in combustion environment: experiments and simulations. *Composite Structures* 127: 77–86. https://doi.org/10.1016/j.compstruct.2015.02.040.

Behrendt, T., Hackemann, S., Mechnich, P. et al. (2016). Development and test of oxide/oxide ceramic matrix composites combustor liner demonstrators for aero-engines. *Journal of Engineering for Gas Turbines and Power* 139: 031705. https://doi.org/10.1115/1.4034515.

Ben Ramdane, C., Julian-Jankowiak, A., Valle, R. et al. (2017). Microstructure and mechanical behaviour of a Nextel™ 610/alumina weak matrix composite subjected to tensile and compressive loadings. *Journal of the European Ceramic Society* 37: 2919–2932. https://doi.org/10.1016/j.jeurceramsoc.2017.02.042.

Bertrand, D.J., Sabelkin, V., Zawada, L., and Mall, S. (2015). Fatigue behavior of sylramic-iBN/BN/CVI SiC ceramic matrix composite in combustion environment. *Journal of Materials Science* 50: 7437–7447. https://doi.org/10.1007/s10853-015-9302-8.

Brennan, J.J. (1990). Interfacial studies of chemical-vapor-infiltrated ceramic matrix composites. *Materials Science and Engineering A* 126: 203–223. https://doi.org/10.1016/0921-5093(90)90126-N.

Bunsell, A.R. and Berger, M.H. (2000). Fine diameter ceramic fibers. *Journal of the European Ceramic Society* 20: 2249–2260. https://doi.org/10.1016/S0955-2219(00)00090-X.

Chawla, N., Holmes, J.W., and Lowden, R.A. (1996). The role of interfacial coatings on the high frequency fatigue behavior of Nicalon/C/SiC composites. *Scripta Materialia* 35: 1411–1416. https://doi.org/10.1016/S1359-6462(96)00326-0.

Corman, G., Upadhyay, R., Sinha, S. et al. (2016). General electric company: selected applications of ceramics and composite materials. In: *Materials Research for Manufacturing: An Industrial Perspective of Turning Materials into New Products*. Cham, Switzerland: Springer International Publishing. https://doi.org/10.1007/978-3-319-23419-9_3.

Davis, J.B., Löfvander, J., Evans, A.G. et al. (1993). Fiber coating concepts for brittle-matrix composites. *Journal of the American Ceramic Society* 76: 1249–1257. https://doi.org/10.1111/j.1151-2916.1993.tb03749.x.

DiCarlo, J.A. and Yun, H.M. (2005). Non-oxide (silicon carbide) fibers. In: *Handbook of Ceramic Composites* (ed. N.P. Bansal). Boston, MA: Springer. https://doi.org/10.1007/0-387-23986-3_2.

Elahi, M., Liao, K., Lesko, J. et al. (1996). Elevated temperature cyclic fatigue of silicon carbide fiber reinforced silicon carbide matrix composites. In: *Ceramic Engineering and Science Proceedings*, vol. 17, 357–361. https://doi.org/10.1002/9780470314500.ch1.

Faber, K.T. (1997). Ceramic composite interfaces: properties and design. *Annual Review of Materials Science* 27: 499–524. https://doi.org/10.1146/annurev.matsci.27.1.499.

Filipuzzi, L. and Naslain, R. (1994). Oxidation mechanisms and kinetics of 1D-SiC/C/SiC composite materials: II. Modeling. *Journal of the American Ceramic Society* 77: 467–480. https://doi.org/10.1111/j.1151-2916.1994.tb07016.x.

Filipuzzi, L., Camus, G., Naslain, R., and Thebault, J. (1994). Oxidation mechanisms and kinetics of 1D-SiC/C/SiC composite materials: I. An experimental approach. *Journal of the American Ceramic Society* 77: 459–466. https://doi.org/10.1111/j.1151-2916.1994.tb07015.x.

Holmquist, M., Lundberg, R., Sudre, O. et al. (2000). Alumina/alumina composite with a porous zirconia interphase – processing, properties and component testing. *Journal of the European Ceramic Society* 20: 599–606. https://doi.org/10.1016/S0955-2219(99)00258-7.

Jacobson, N., Farmer, S., Moore, A., and Sayir, H. (1999). High-temperature oxidation of boron nitride: I. Monolithic boron nitride. *Journal of the American Ceramic Society* 82: 393–398. https://doi.org/10.1111/j.1551-2916.1999.tb20075.x.

Kaneko, Y., Zhu, S., Ochi, Y. et al. (2001). Effect of frequency on fatigue behavior in Tyranno fiber reinforced SiC composites. In: *Ceramic Engineering and Science Proceedings*, vol. 22, 553–560. https://doi.org/10.1002/9780470294680.ch64.

Keller, K.A., Mah, T.I., Parthasarathy, T.A., and Cooke, C.M. (2000). Fugitive interfacial carbon coatings for oxide/oxide composites. *Journal of the American Ceramic Society* 83: 329–336. https://doi.org/10.1111/j.1151-2916.2000.tb01194.x.

Keller, K.A., Mah, T.I., Parthasarathy, T.A. et al. (2003). Effectiveness of monazite coatings in oxide/oxide composites after long-term exposure at high temperature. *Journal of the American Ceramic Society* 86: 325–332. https://doi.org/10.1111/j.1151-2916.2003.tb00018.x.

Kerans, R.J. (1996). Viability of oxide fiber coatings in ceramic composites for accommodation of misfit stresses. *Journal of the American Ceramic Society* 79: 1664–1668. https://doi.org/10.1111/j.1151-2916.1996.tb08779.x.

Kim, T.T., Mall, S., Zawada, L.P., and Jefferson, G. (2010). Simultaneous fatigue and combustion exposure of a SiC/SiC ceramic matrix composite. *Journal of Composite Materials* 44: 2991–3016. https://doi.org/10.1177/0021998310373519.

Kiser, J.D., Bansal, N.P., Szelagowski, J., et al. (2015). Oxide/oxide ceramic matrix composite (CMC) exhaust mixer development in the NASA environmentally responsible aviation (ERA) project. *Proceedings of ASME Turbo Expo 2015: Turbine Technical Conference and Exposition*, Montreal, Quebec, Canada (15–19 June 2015). New York: ASME.

Krenkel, W. (ed.) (2008). *Ceramic Matrix Composites: Fiber-Reinforced Ceramics and Their Applications*. Weinheim: Wiley-VCH. https://doi.org/10.1002/9783527622412.

Kuntz, M., Meier, B., and Grathwohl, G. (1993). Residual stresses in fiber-reinforced ceramics due to thermal expansion mismatch. *Journal of the American Ceramic Society* 76: 2607–2612. https://doi.org/10.1111/j.1151-2916.1993.tb03988.x.

Lanser, R.L. and Ruggles-Wrenn, M.B. (2016). Tension-compression fatigue of a Nextel™ 720/alumina composite at 1200 °C in air and in steam. *Applied Composite Materials* 23: 707–717. https://doi.org/10.1007/s10443-016-9481-8.

Lebel, L., Turenne, S., and Boukhili, R. (2017). An experimental apparatus and procedure for the simulation of thermal stresses in gas turbine combustion chamber panels made of ceramic matrix composites. *Journal of Engineering for Gas Turbines and Power* 139: 091502. https://doi.org/10.1115/1.4035906.

Li, L. (2018). *Damage, Fracture and Fatigue of Ceramic-Matrix Composites*. Springer Nature Singapore Private Limited. ISBN: 978-981-13-1782-8. https://doi.org/10.1007/978-981-13-1783-5.

Li, L. (2019). *Thermomechanical Fatigue of Ceramic-Matrix Composites*. Wiley-VCH. ISBN: 978-3-527-34637-0. https://onlinelibrary.wiley.com/doi/book/10.1002/9783527822614.

Liu, H., Yang, J., Zhou, Y. et al. (2018). Progress in coupon tests of SiC_f/SiC ceramic matrix composites used for aero engines. *Journal of Materials Engineering* 46: 1–12.

Luthra, K.L. (1994). Theoretical aspects of the oxidation of silica-forming ceramics. In: *Corrosion of Advanced Ceramics*, NATO Science Series E: (Closed), vol. 267 (ed. K.G. Nickel). Dordrecht: Springer. https://doi.org/10.1007/978-94-011-1182-9_2.

Mall, S. and Ahn, J.M. (2008). Frequency effects on fatigue behavior of Nextel 720™/alumina at room temperature. *Journal of the European Ceramic Society* 28: 2783–2789. https://doi.org/10.1016/j.jeurceramsoc.2008.04.005.

Mizuno, M., Zhu, S., Nagano, Y. et al. (1996). Cyclic-fatigue behavior of SiC/SiC composites at room and high temperatures. *Journal of the American Ceramic Society* 79: 3065–3077. https://doi.org/10.1111/j.1151-2916.1996.tb08078.x.

Morgan, P.E.D. and Marshall, D.B. (1995). Ceramic composites of monazite and alumina. *Journal of the American Ceramic Society* 78: 1553–1563. https://doi.org/10.1111/j.1151-2916.1995.tb08851.x.

Morscher, G.N. (2010). Tensile creep and rupture of 2D-woven SiC/SiC composites for high temperature applications. *Journal of the European Ceramic Society* 30: 2209–2221. https://doi.org/10.1016/j.jeurceramsoc.2010.01.030.

Morscher, G.N., Hurst, J., and Brewer, D. (2000). Intermediate-temperature stress rupture of a woven Hi-Nicalon, BN-interphase, SiC-matrix composite in air. *Journal of the American Ceramic Society* 83: 1441–1449. https://doi.org/10.1111/j.1151-2916.2000.tb01408.x.

Morscher, G.N., Ojard, G., Miller, R. et al. (2008). Tensile creep and fatigue of Sylramic-iBN melt-infiltrated SiC matrix composites: retained properties, damage development, and failure mechanisms. *Composites Science and Technology* 68: 3305–3313. https://doi.org/10.1016/j.compscitech.2008.08.028.

Nasiri, N.A., Patra, N., Ni, N. et al. (2016). Oxidation behaviour of SiC/SiC ceramic matrix composites in air. *Journal of the European Ceramic Society* 36: 3293–3302. https://doi.org/10.1016/j.jeurceramsoc.2016.05.051.

Reynaud, P. (1996). Cyclic fatigue of ceramic-matrix composites at ambient and elevated temperatures. *Composites Science and Technology* 56: 809–814. https://doi.org/10.1016/0266-3538(96)00025-5.

Reynaud, P., Rouby, D., and Fantozzi, G. (1994). Effects of interfacial evolutions on the mechanical behaviour of ceramic matrix composites during cyclic fatigue. *Scripta Metallurgica et Materialia* 31: 1061–1066. https://doi.org/10.1016/0956-716X(94)90527-4.

van Roode, M. and Bhattacharya, A.K. (2013). Durability of oxide/oxide ceramic matrix composites in gas turbine combustors. *Journal of Engineering for Gas Turbines and Power* 135: 051301. https://doi.org/10.1115/1.4007978.

Rouby, D. and Reynaud, P. (1993). Fatigue behaviour related to interface modification during load cycling in ceramic-matrix fibre composites. *Composites Science and Technology* 48: 109–118. https://doi.org/10.1016/0266-3538(93)90126-2.

Roy, J., Chandra, S., Das, S., and Maitra, S. (2014). Oxidation behaviour of silicon carbide – a review. *Reviews on Advanced Materials Science* 38: 29–39.

Ruggles-Wrenn, M.B. and Braun, J.C. (2008). Effects of steam environment on creep behavior of Nextel™ 720/alumina ceramic composite at elevated temperature. *Materials Science and Engineering A* 497: 101–110. https://doi.org/10.1016/j.msea.2008.06.036.

Ruggles-Wrenn, M.B. and Laffey, P. (2008). Creep behavior in interlaminar shear of Nextel™ 720/alumina ceramic composite at elevated temperature in air and in steam. *Composites Science and Technology* 68: 2260–2266. https://doi.org/10.1016/j.compscitech.2008.04.009.

Ruggles-Wrenn, M.B. and Lanser, R.L. (2016). Tension-compression fatigue of an oxide/oxide ceramic composite at elevated temperature. *Materials Science and Engineering A* 659: 270–277. https://doi.org/10.1016/j.msea.2016.02.057.

Ruggles-Wrenn, M.B. and Szymczak, N.R. (2008). Effects of steam environment on compressive creep behavior of Nextel™ 720/alumina ceramic composite at 1200 °C. *Composites Part A Applied Science and Manufacturing* 39: 1829–1837. https://doi.org/10.1016/j.compositesa.2008.09.005.

Ruggles-Wrenn, M.B. and Whiting, B.A. (2011). Cyclic creep and recovery behavior of Nextel™ 720/alumina ceramic composite at 1200 °C. *Materials Science and Engineering A* 528: 1848–1856. https://doi.org/10.1016/j.msea.2010.10.011.

Ruggles-Wrenn, M.B., Mall, S., Eber, C.A., and Harlan, L.B. (2006). Effects of steam environment on high-temperature mechanical behavior of Nextel™ 720/alumina (N720/A) continuous fiber ceramic composite. *Composites Part A Applied Science and Manufacturing* 37: 2029–2040. https://doi.org/10.1016/j.compositesa.2005.12.008.

Ruggles-Wrenn, M.B., Hetrick, G., and Baek, S.S. (2008a). Effects of frequency and environment on fatigue behavior of an oxide–oxide ceramic composite at 1200 °C. *International Journal of Fatigue* 30: 502–516. https://doi.org/10.1016/j.ijfatigue.2007.04.004.

Ruggles-Wrenn, M.B., Radzicki, A.T., Baek, S.S., and Keller, K.A. (2008b). Effect of loading rate on the monotonic tensile behavior and tensile strength of an oxide–oxide ceramic composite at 1200 °C. *Materials Science and Engineering A* 492: 88–94. https://doi.org/10.1016/j.msea.2008.03.006.

Ruggles-Wrenn, M.B., Siegert, G.T., and Baek, S.S. (2008c). Creep behavior of Nextel™ 720/alumina ceramic composite with ±45° fiber orientation at 1200 °C.

Composites Science and Technology 68: 1588–1595. https://doi.org/10.1016/j.compscitech.2007.07.012.

Ruggles-Wrenn, M.B., Christensen, D.T., Chamberlain, A.L. et al. (2011). Effect of frequency and environment on fatigue behavior of a CVI SiC/SiC ceramic matrix composite at 1200 °C. *Composites Science and Technology* 71: 190–196. https://doi.org/10.1016/j.compscitech.2010.11.008.

Ruggles-Wrenn, M.B., Boucher, N., and Przybyla, C. (2018). Fatigue of three advanced SiC/SiC ceramic matrix composites at 1200 °C in air and in steam. *International Journal of Applied Ceramic Technology* 15: 3–15.

Sabelkin, V., Mall, S., Cook, T.S., and Fish, J. (2016). Fatigue and creep behaviors of a SiC/SiC composite under combustion and laboratory environments. *Journal of Composite Materials* 50: 2145–2153. https://doi.org/10.1177/0021998315602323.

Singh, A.K., Sabelkin, V., and Mall, S. (2017a). Fatigue behavior of double-edge notched oxide/oxide ceramic matrix composite in a combustion environment. *Journal of Composite Materials* 51: 3669–3683. https://doi.org/10.1177/0021998317692655.

Singh, A.K., Sabelkin, V., and Mall, S. (2017b). Creep-rupture behaviour of notched oxide/oxide ceramic matrix composite in combustion environment. *Advances in Applied Ceramics* 117: 30–41. https://doi.org/10.1080/17436753.2017.1359444.

Steel, S.G., Zawada, L.P., and Mall, S. (2001). Fatigue behavior of a Nextel™ 720/alumina (N720/A) composite at room and elevated temperature. In: *Ceramic Engineering and Science Proceedings*, vol. 22, 695–702. https://doi.org/10.1002/9780470294680.ch80.

Ünal, Ö. (1996a). Cyclic tensile stress-strain behaviour of SiC/SiC composites. *Journal of Materials Science Letters* 15: 789–791. https://doi.org/10.1007/BF00274605.

Ünal, Ö. (1996b). Tensile and fatigue behavior of a SiC/SiC composite at 1300 °C. In: *Symposium on Thermal and Mechanical Test Methods and Behavior of Continuous-Fiber Ceramic Composites* (ASTM-STP-1309). Philadelphia: American Society for Testing and Materials. https://doi.org/10.2172/238543.

Ünal, Ö. (1996c). Low-cycle tensile fatigue behavior of a SiC/SiC composite. In: *Ceramic Engineering and Science Proceedings*, vol. 17, 157–165. https://doi.org/10.1002/9780470314876.ch16.

Ünal, Ö., Eckel, A.J., and Laabs, F.C. (1995). The 1400 °C-oxidation effect on microstructure, strength and cyclic life of SiC/SiC composites. *Scripta Metallurgica et Materialia* 33: 983–988. https://doi.org/10.1016/0956-716X(95)00309-J.

Wilson, D.M. and Visser, L.R. (2001). High performance oxide fibers for metal and ceramic composites. *Composites Part A Applied Science and Manufacturing* 32: 1143–1153. https://doi.org/10.1016/S1359-835X(00)00176-7.

Yang, R., Qi, Z., Yang, J., and Jiao, J. (2018). Research progress in oxide/oxide ceramic matrix composites and processing technologies. *Journal of Materials Engineering* 46: 1–9.

Zawada, L.P., Hay, R.S., Lee, S.S., and Staehler, J. (2003). Characterization and high-temperature mechanical behavior of an oxide/oxide composite. *Journal of the American Ceramic Society* 86: 981–990. https://doi.org/10.1111/j.1151-2916.2003.tb03406.x.

Zhu, S. (2006). Fatigue behavior of ceramic matrix composite oxidized at intermediate temperatures. *Materials Transactions* 47: 1965–1967. https://doi.org/10.2320/matertrans.47.1965.

Zhu, S. and Kagawa, Y. (2001). Fatigue fracture in SiC fiber reinforced SiC composites. *Seisan Kenkyu* 53: 40–43. https://doi.org/10.11188/seisankenkyu.53.470.

Zhu, S., Kagawa, Y., Mizuno, M. et al. (1996). In situ observation of cyclic fatigue crack propagation of SiC-fiber/SiC composite at room temperature. *Materials Science and Engineering A* 220: 100–107. https://doi.org/10.1016/S0921-5093(96)10440-8.

Zhu, S., Mizuno, M., Kagawa, Y. et al. (1997). Creep and fatigue behavior of SiC fiber reinforced SiC composite at high temperatures. *Materials Science and Engineering A* 225: 69–77. https://doi.org/10.1016/S0921-5093(96)10872-8.

Zhu, S., Mizuno, M., Nagano, Y. et al. (1998). Creep and fatigue behavior of enhanced SiC/SiC composite at high temperatures. *Journal of the American Ceramic Society* 81: 2269–2277. https://doi.org/10.1111/j.1151-2916.1998.tb02621.x.

Zhu, S., Mizuno, M., Kagawa, Y., and Mutoh, Y. (1999). Monotonic tension, fatigue and creep behavior of SiC-fiber-reinforced SiC-matrix composites: a review. *Composites Science and Technology* 59: 833–851. https://doi.org/10.1016/S0266-3538(99)00014-7.

Zhu, S., Kaneko, Y., Ochi, Y. et al. (2002). Effect of frequency on fatigue behavior in 3D-woven Tyranno fiber reinforced SiC composites. *Journal of the Society of Materials Science Japan* 51: 1400–1404. https://doi.org/10.2472/jsms.51.1400.

Zhu, S., Kaneko, Y., Ochi, Y. et al. (2004). Low cycle fatigue behavior in an orthogonal three-dimensional woven Tyranno fiber reinforced Si–Ti–C–O matrix composite. *International Journal of Fatigue* 26: 1069–1074. https://doi.org/10.1016/j.ijfatigue.2004.03.001.

Zok, F.W. (2006). Developments in oxide fiber composites. *Journal of the American Ceramic Society* 89: 3309–3324. https://doi.org/10.1111/j.1551-2916.2006.01342.x.

2

Interface Characterization of Ceramic-Matrix Composites

2.1 Introduction

Ceramic-matrix composites (CMCs) are widely used in the high-temperature field as a light and high-performance structural composite material. High-quality and high-temperature properties make it possible to replace superalloy materials as one of the candidate materials for aero-engines, especially for aero-engine core engines. In the research and application of CMCs, the existing mature aero-engines are fully utilized for assessment and verification from low temperature to high temperature, stator components to rotor components (Naslain 2004; DiCarlo and Roode 2006; DiCarlo et al. 2005; Zok 2016; Lino Alves et al. 2016; Li 2016, 2018a,b, 2020a,b). Firstly, the stator parts with medium temperature (700–1000 °C) and medium load (less than 120 MPa) were developed, i.e. exhaust nozzle flaps and sealings; then the medium stator parts with high temperature (1000–1300 °C) were developed, such as combustion chamber flame tube, flame stabilizer, turbine guide vane, and turbine outer ring; and the stator or rotor parts with higher load (more than 120 MPa), such as high-pressure turbine rotor and stator, have been developed (Naslain 2005; Halbig et al. 2013; Ding 2014). The commercial aero engines require low fuel consumption, low noise, and low NO_x emissions, so new requirements are put forward for pressure ratio and turbine front temperature. CMC has the characteristics of light-weight, high-temperature resistance, corrosion resistance, and impact resistance, so it is expected to be used in combustion chambers, turbines, exhaust nozzles, and other components of the next generation of commercial aero engines (Watanabe and Manabe 2018; Kumar et al. 2018; Newton et al. 2018).

For fiber-reinforced CMCs, the mechanical properties depend tremendously on the load transfer at the fiber/matrix interface (Rebillat 2014; Xia and Li 2014). The interface properties of the fiber/matrix interface shear stress and the interface debonding energy affect the tensile and fatigue behavior of fiber-reinforced CMCs (Li et al. 2013, 2015; Li 2018a, 2019a,b,c). Vagaggini et al. (1995) developed an approach to establish the relationship between the interface properties and the hysteresis loops of fiber-reinforced CMCs and divided the interface debonding energy into small and large, which affects the shape of the hysteresis loops upon unloading and reloading. Domergue et al. (1995) measured the interface properties of unidirectional SiC/CAS (silicon

Interfaces of Ceramic-Matrix Composites: Design, Characterization and Damage Effects,
First Edition. Longbiao Li. © 2020 WILEY-VCH GmbH.

carbide/calcium aluminosilicate) and SiC/SiC composites using the hysteresis loops, and the interface shear stress of SiC/SiC composite is much higher than that of SiC/CAS composite, leading to un-saturation of matrix cracking of SiC/SiC composite till tensile fracture. Curtin et al. (1998) predicted the tensile stress–strain behavior of mini SiC/SiC composite considering matrix cracking evolution, fiber damage, and ultimate failure. It was found that the matrix cracking stress affects the brittle and tough behavior of fiber-reinforced CMCs. Carrere et al. (2000) investigated the influence of the interphase on the matrix cracking deflection in mini SiC/C/SiC composite with a pyrocarbon interphase. The deflection of the matrix cracking depends on the interface bond strength between the fiber and the matrix. Xia and Curtin (2000) investigated the high interface shear stress on the tensile strength of fiber-reinforced CMCs considering the stress concentration at the interface debonding tip. When the bond strength is high and the fiber/matrix interface frictional shear stress in the debonding region is low, the stress concentration occurs near the interface debonding tip. Sauder et al. (2010) investigated the influence of the interface characteristics on the tensile and loading/unloading behavior of two different mini SiC/SiC composites. The interphase thickness and the fiber surface roughness affect the fiber/matrix interface shear stress at the debonding region, and then the tensile behavior of fiber-reinforced CMCs. Under cyclic loading, the fiber/matrix interface shear stress decreases with applied cycles, which depends on the peak stress, stress ratio, loading frequency, temperature, and environment (Cho et al. 1991; Holmes and Cho 1992; Evans et al. 1995; Reynaud 1996; Zhu et al. 1999; Staehler et al. 2003; Fantozzi and Reynaud 2009; Ruggles-Wrenn et al. 2011; Dassios et al. 2013).

In this chapter, the effect of the fiber/matrix interface properties and pre-exposure on the tensile and fatigue behavior of fiber-reinforced CMCs is investigated. The relationships between the interface properties and the composite tensile and fatigue damage are established. The effects of the interface properties and pre-exposure temperature and time on the first matrix cracking stress, matrix cracking evolution, first and complete interface debonding stress, fatigue hysteresis dissipated energy, fatigue hysteresis modulus, and fatigue hysteresis width, and tensile damage and fracture processes are analyzed. The fatigue life of fiber-reinforced CMCs with different fiber preforms, i.e. unidirectional, cross-ply, 2D, 2.5D, and 3D CMCs at room and elevated temperatures in air and oxidative environment, are predicted using the micromechanics approach. The experimental tensile and fatigue damage and fatigue life of different CMCs are predicted for different interface properties and testing conditions.

2.2 Effect of Interface Properties on Tensile and Fatigue Behavior of Ceramic-Matrix Composites

In this section, the effect of the fiber/matrix interface properties on the tensile and fatigue behavior of 2D SiC/SiC composites is investigated. The relationships between the interface properties and the composite tensile and fatigue damage

2.2 Effect of Interface Properties on Tensile and Fatigue Behavior

parameters are established. The effects of the interface properties on the first matrix cracking stress, matrix cracking evolution, first and complete interface debonding stress, fatigue hysteresis dissipated energy, fatigue hysteresis modulus, and fatigue hysteresis width are analyzed. The experimental tensile and fatigue behavior of SiC/SiC composites are predicted for different interface properties.

2.2.1 Theoretical Analysis

In the present analysis, the fiber failure is considered in the analysis of the first matrix cracking stress, matrix cracking density, interface debonding stress, and fatigue hysteresis-based damage parameters.

2.2.1.1 First Matrix Cracking Stress

For the first matrix cracking of fiber-reinforced CMCs, the energy balance relationship can be determined as

$$\alpha \sigma^2 + \beta \sigma + \gamma = 0 \tag{2.1}$$

where

$$\alpha = \frac{l_d}{E_c} \tag{2.2}$$

$$\beta = -\frac{2 V_f l_d T}{E_c} \tag{2.3}$$

$$\gamma = \frac{4}{3} \left(\frac{\tau_i}{r_f}\right)^2 \left(\frac{V_f E_c}{V_m E_m E_f}\right) l_d^3 + \frac{V_f l_d T^2}{E_f} - \frac{2 V_f \tau_i l_d^2}{r_f E_f} T - V_m \zeta_m - \frac{4 V_f l_d}{r_f} \zeta_d \tag{2.4}$$

where l_d denotes the fiber/matrix interface debonding length; E_f, E_m, and E_c denote the fiber, matrix, and composite elastic modulus, respectively; V_f and V_m denote the fiber and the matrix volume, respectively; r_f denotes the fiber radius; ζ_m and ζ_d denote the matrix fracture energy and the interface debonding energy, respectively; τ_i denotes the fiber/matrix interface shear stress in the debonding region; and Φ denotes the fiber intact stress.

$$\frac{\sigma}{V_f} = \Phi \left(\frac{\sigma_c}{\Phi}\right)^{m_f+1} \left\{1 - \exp\left[-\left(\frac{\Phi}{\sigma_c}\right)^{m_f+1}\right]\right\} \tag{2.5}$$

where σ_c denotes the fiber characteristic strength; and m_f denotes the fiber Weibull modulus.

2.2.1.2 Matrix Cracking Density

The energy balance relationship to evaluate the matrix cracking evolution is given by:

$$U_m(\sigma > \sigma_{mc}, l_c, l_d) = U_{crm}(\sigma_{mc}, l_0) \tag{2.6}$$

where

$$U_m = \begin{cases} \dfrac{A_m}{E_m}\left\{\dfrac{4}{3}\left(\dfrac{V_f\tau_i}{V_m r_f}l_d\right)^2 l_d + \sigma_{mo}^2\left(\dfrac{l_c}{2}-l_d\right)\right. \\ \left. -2\sigma_{mo}\left[\dfrac{V_f}{V_m}(\Phi-\sigma_{fo})-2\dfrac{V_f\tau_i}{r_f V_m}l_d\right]\left(-\dfrac{r_f}{\rho}\right)\left[\exp\left(-\rho\dfrac{l_c/2-l_d}{r_f}\right)-1\right] \right. \\ \left. +\left[\dfrac{V_f}{V_m}(\Phi-\sigma_{fo})-2\dfrac{V_f\tau_i}{r_f V_m}l_d\right]^2\left(-\dfrac{r_f}{2\rho}\right)\left[\exp\left(-2\rho\dfrac{l_c/2-l_d}{r_f}\right)-1\right]\right\}, & l_d < \dfrac{l_c}{2} \\ \dfrac{A_m l_c^3}{6E_m}\left(\dfrac{\tau_i V_f}{r_f V_m}\right)^2, & l_d = \dfrac{l_c}{2} \end{cases}$$

(2.7)

$$U_{crm} = \dfrac{1}{2}kA_m l_0 \dfrac{\sigma_{mocr}^2}{E_m} \qquad (2.8)$$

where A_m is the cross-section area of matrix in the unit cell; k denotes the critical matrix strain energy parameter; l_0 is the initial matrix crack spacing; and σ_{mocr} denotes the matrix axial stress in the interface bonded region at the first matrix cracking stress.

2.2.1.3 Fatigue Hysteresis-Based Damage Parameters

The initial fiber/matrix interface debonding stress σ_d and the interface complete debonding stress σ_b can be obtained as

$$\sigma_d = \dfrac{V_f E_c \tau_i}{\rho V_m E_m}\left[1+\sqrt{1+4\dfrac{V_m E_m E_f}{r_f E_c}\dfrac{\rho^2}{\tau_i^2}\zeta_d}\right] \qquad (2.9)$$

$$\sigma_b = \dfrac{V_f E_c \tau_i}{\rho V_m E_m}\left[1+\rho\dfrac{l_c}{r_f}+\sqrt{1+4\dfrac{V_m E_m E_f}{r_f E_c}\dfrac{\rho^2}{\tau_i^2}\zeta_d}\right] \qquad (2.10)$$

The fatigue hysteresis dissipated energy U_e can be given by:

$$U_e = \int_{\sigma_{min}}^{\sigma_{max}}[\varepsilon_{unloading}(\sigma)-\varepsilon_{reloading}(\sigma)]\,d\sigma \qquad (2.11)$$

where

$$\varepsilon_{unloading} = \begin{cases} \dfrac{\Phi_U}{E_f}+4\dfrac{\tau_i}{E_f}\dfrac{y^2}{r_f l_c}-\dfrac{\tau_i}{E_f}\dfrac{(2y-l_d)(2y+l_d-l_c)}{r_f l_c}-(\alpha_c-\alpha_f)\Delta T, & l_d < \dfrac{l_c}{2} \\ \dfrac{\Phi_U}{E_f}+4\dfrac{\tau_i}{E_f}\dfrac{y^2}{r_f l_c}-2\dfrac{\tau_i}{E_f}\dfrac{(2y-l_c/2)^2}{r_f l_c}-(\alpha_c-\alpha_f)\Delta T, & l_d = \dfrac{l_c}{2} \end{cases}$$

(2.12)

$$\varepsilon_{reloading} = \begin{cases} \dfrac{\Phi_R}{E_f} - 4\dfrac{\tau_i}{E_f}\dfrac{z^2}{r_f l_c} + 4\dfrac{\tau_i}{E_f}\dfrac{(y-2z)^2}{r_f l_c} \\ \quad + 2\dfrac{\tau_i}{E_f}\dfrac{(l_d - 2y + 2z)(l_d + 2y - 2z - l_c)}{r_f l_c} - (\alpha_c - \alpha_f)\Delta T, \quad l_d < \dfrac{l_c}{2} \\ \dfrac{\Phi_R}{E_f} - 4\dfrac{\tau_i}{E_f}\dfrac{z^2}{r_f l_c} + 4\dfrac{\tau_i}{E_f}\dfrac{(y-2z)^2}{r_f l_c} \\ \quad - 2\dfrac{\tau_i}{E_f}\dfrac{(l_c/2 - 2y + 2z)^2}{r_f l_c} - (\alpha_c - \alpha_f)\Delta T, \quad l_d = \dfrac{l_c}{2} \end{cases}$$

(2.13)

where Φ_U and Φ_R denote the intact fiber stress upon unloading and reloading, respectively.

The fatigue hysteresis width $\Delta\varepsilon$ can be given by:

$$\Delta\varepsilon = \varepsilon_{unloading}\left(\dfrac{\sigma_{min} + \sigma_{max}}{2}\right) - \varepsilon_{reloading}\left(\dfrac{\sigma_{min} + \sigma_{max}}{2}\right) \quad (2.14)$$

The fatigue hysteresis modulus E can be obtained as

$$E = \dfrac{\sigma_{max} - \sigma_{min}}{\varepsilon_{max}(\sigma_{max}) - \varepsilon_{min}(\sigma_{min})} \quad (2.15)$$

2.2.2 Results and Discussion

The effects of fiber/matrix interface properties on the tensile and fatigue damage are analyzed. The SiC/SiC composite is used to the case analysis, and the material properties are given by $V_f = 15\%$, $E_f = 400\,\text{GPa}$, $E_m = 350\,\text{GPa}$, $r_f = 6\,\mu m$, $\alpha_f = 4.5\times 10^{-6}/°C$, $\alpha_m = 4.6\times 10^{-6}/°C$, $\Delta T = -1000\,°C$, $\zeta_m = 6\,\text{J/m}^2$, $\zeta_d = 1.2\,\text{J/m}^2$, $\tau_i = 30\,\text{MPa}$, $\sigma_c = 3.0\,\text{GPa}$, and $m_f = 5$.

2.2.2.1 Effect of the Interface Properties on First Matrix Cracking Stress

The first matrix cracking stress, interface debonding length, and broken fibers fraction versus the interface shear stress and interface debonding energy curves for different fiber volume are shown in Figures 2.1 and 2.2. When the interface shear stress and interface debonding energy increase, the first matrix cracking stress increases, the fiber/matrix interface debonding length decreases, and the broken fibers fraction increases.

When the fiber volume is $V_f = 15\%$, the first matrix cracking stress increases from $\sigma_{mc} = 132\,\text{MPa}$ at the interface shear stress of $\tau_i = 10\,\text{MPa}$ to $\sigma_{mc} = 180\,\text{MPa}$ at the interface shear stress of $\tau_i = 50\,\text{MPa}$, and from $\sigma_{mc} = 143\,\text{MPa}$ at the interface debonding energy of $\zeta_d/\zeta_m = 0.1$ to $\sigma_{mc} = 242\,\text{MPa}$ at the interface debonding energy of $\zeta_d/\zeta_m = 0.9$; the interface debonding length decreases from $l_d/r_f = 10.8$ at the interface shear stress of $\tau_i = 10\,\text{MPa}$ to $l_d/r_f = 4.9$ at the interface shear stress of $\tau_i = 50\,\text{MPa}$, and from $l_d/r_f = 7.2$ at the interface debonding energy of $\zeta_d/\zeta_m = 0.1$ to $l_d/r_f = 4.7$ at the interface debonding energy of $\zeta_d/\zeta_m = 0.9$; and the broken fibers fraction increases from $q = 0.06\%$ at the interface shear stress of $\tau_i = 10\,\text{MPa}$ to $q = 0.42\%$ at the interface shear stress of $\tau_i = 50\,\text{MPa}$, and from

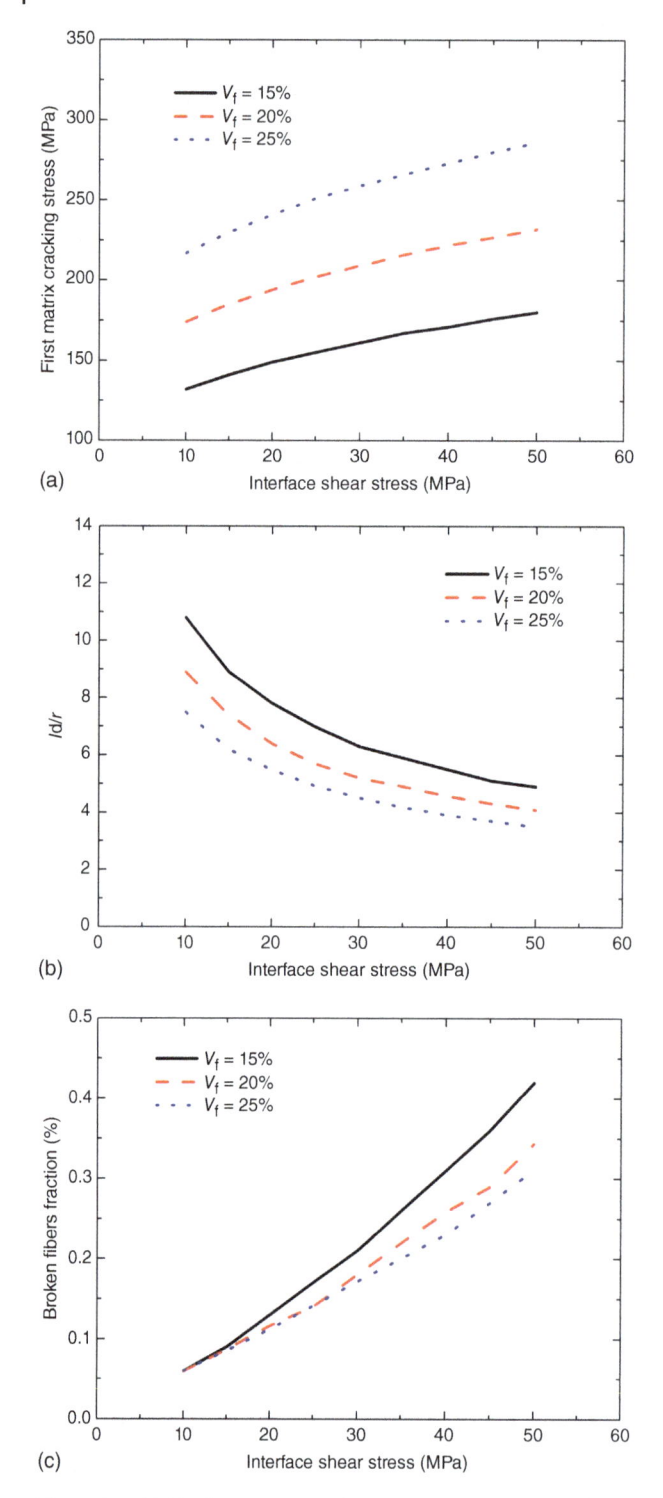

Figure 2.1 (a) The first matrix cracking stress versus the interface shear stress curves; (b) the interface debonding length versus the interface shear stress curves; and (c) the broken fibers fraction versus the interface shear stress curves for different fiber volume.

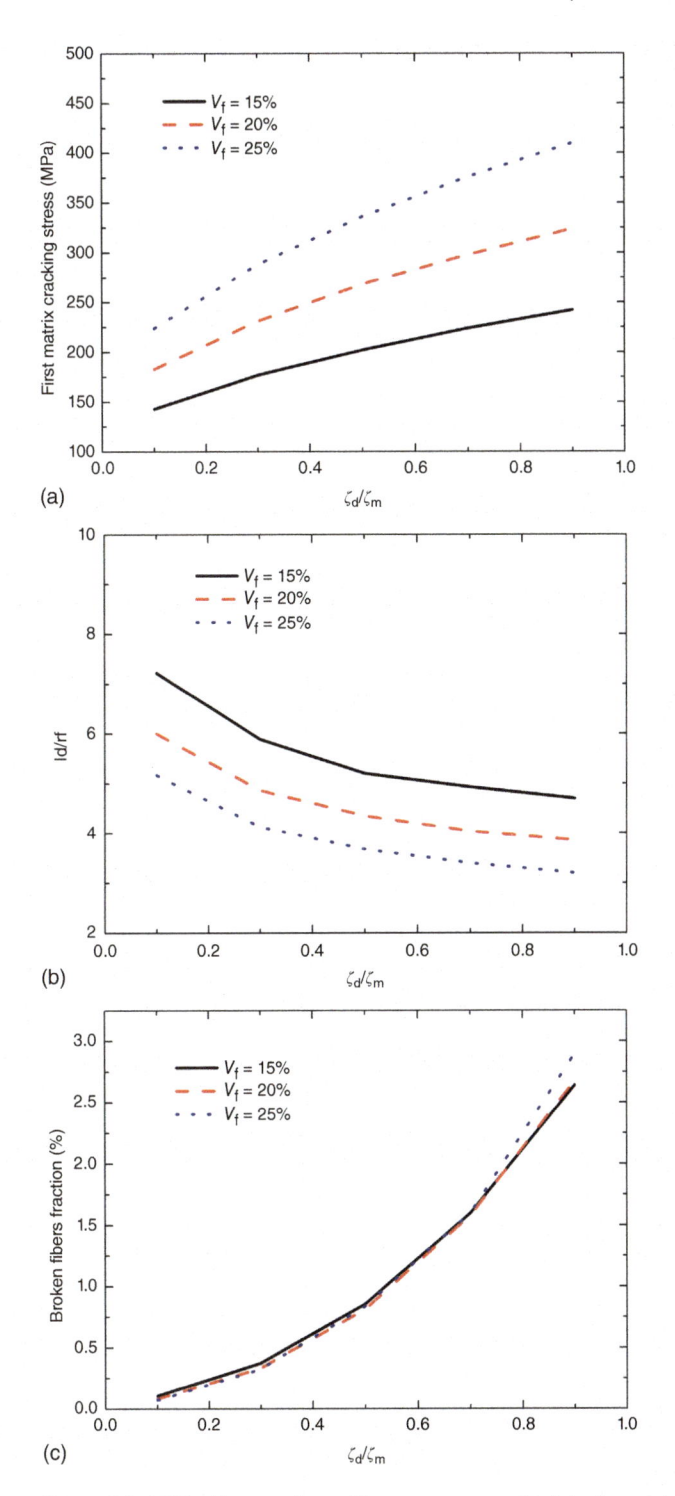

2.2 Effect of Interface Properties on Tensile and Fatigue Behavior | 35

Figure 2.2 (a) The first matrix cracking stress versus the interface debonding energy curves; (b) the interface debonding length versus the interface debonding energy curves; and (c) the broken fibers fraction versus the interface debonding energy curves for different fiber volume.

$q = 0.1\%$ at the interface debonding energy of $\zeta_d/\zeta_m = 0.1$ to $q = 2.6\%$ at the interface debonding energy of $\zeta_d/\zeta_m = 0.9$.

When the fiber volume is $V_f = 25\%$, the first matrix cracking stress increases from $\sigma_{mc} = 217$ MPa at the interface shear stress of $\tau_i = 10$ MPa to $\sigma_{mc} = 286$ MPa at the interface shear stress of $\tau_i = 50$ MPa, and from $\sigma_{mc} = 224$ MPa at the interface debonding energy of $\zeta_d/\zeta_m = 0.1$ to $\sigma_{mc} = 410$ MPa at the interface debonding energy of $\zeta_d/\zeta_m = 0.9$; the fiber/matrix interface debonding length decreases from $l_d/r_f = 7.5$ at the interface shear stress of $\tau_i = 10$ MPa to $l_d/r_f = 3.5$ at the interface shear stress of $\tau_i = 50$ MPa, and from $l_d/r_f = 5.1$ at the interface debonding energy of $\zeta_d/\zeta_m = 0.1$ to $l_d/r_f = 3.2$ at the interface debonding energy of $\zeta_d/\zeta_m = 0.9$; and the broken fibers fraction increases from $q = 0.06\%$ at the interface shear stress of $\tau_i = 10$ MPa to $q = 0.31\%$ at the interface shear stress of $\tau_i = 50$ MPa, and from $q = 0.07\%$ at the interface debonding energy of $\zeta_d/\zeta_m = 0.1$ to $q = 2.9\%$ at the interface debonding energy of $\zeta_d/\zeta_m = 0.9$.

2.2.2.2 Effect of the Interface Properties on Matrix Cracking Density

The matrix cracking density, interface debonding length, and broken fibers fraction versus the applied stress curves for different interface shear stress and interface debonding energy are shown in Figures 2.3 and 2.4. When the interface shear stress increases, the matrix cracking density, saturation matrix cracking stress, and interface debonding length increase; and when the interface debonding energy increases, the matrix cracking density decreases, and the saturation matrix cracking stress increases.

When the interface shear stress is $\tau_i = 20$ MPa, matrix cracking density increases from $\lambda = 0.09$/mm at the first matrix cracking stress of $\sigma_{mc} = 155$ MPa to $\lambda = 2.5$/mm at the saturation matrix cracking stress of $\sigma_{sat} = 230$ MPa; the interface debonding length increases from $2l_d/l_c = 0.9\%$ at the first matrix cracking stress of $\sigma_{mc} = 155$ MPa to $2l_d/l_c = 60.8\%$ at the saturation matrix cracking stress of $\sigma_{sat} = 230$ MPa; and the broken fibers fraction increases from $q = 0.169\%$ at the first matrix cracking stress of $\sigma_{mc} = 155$ MPa to $q = 2\%$ at the saturation matrix cracking stress of $\sigma_{sat} = 230$ MPa. When the interface shear stress is $\tau_i = 40$ MPa, matrix cracking density increases from $\lambda = 0.13$/mm at the first matrix cracking stress of $\sigma_{mc} = 178$ MPa to $\lambda = 3.9$/mm at the saturation matrix cracking stress of $\sigma_{sat} = 267$ MPa; the interface debonding length increases from $2l_d/l_c = 0.96\%$ at the first matrix cracking stress of $\sigma_{mc} = 178$ MPa to $2l_d/l_c = 63.4\%$ at the saturation matrix cracking stress of $\sigma_{sat} = 267$ MPa; and the broken fibers fraction increases from $q = 0.38\%$ at the first matrix cracking stress of $\sigma_{mc} = 178$ MPa to $q = 4.9\%$ at the saturation matrix cracking stress of $\sigma_{sat} = 267$ MPa.

When the interface debonding energy is $\zeta_d/\zeta_m = 0.1$, the matrix cracking density increases from $\lambda = 0.1$/mm at the first matrix cracking stress of $\sigma_{mc} = 149$ MPa to $\lambda = 3.5$/mm at the saturation matrix cracking stress of $\sigma_{sat} = 240$ MPa; the interface debonding length increases from $2l_d/l_c = 1\%$ at the first matrix cracking stress of $\sigma_{mc} = 149$ MPa to $2l_d/l_c = 72\%$ at the saturation matrix cracking stress of $\sigma_{sat} = 240$ MPa; and the broken fibers fraction increases from $q = 0.1\%$ at the first matrix cracking stress of $\sigma_{mc} = 149$ MPa to $q = 2.4\%$ at the saturation matrix cracking stress of $\sigma_{sat} = 240$ MPa. When the interface debonding energy is $\zeta_d/\zeta_m = 0.5$, the matrix cracking density increases from

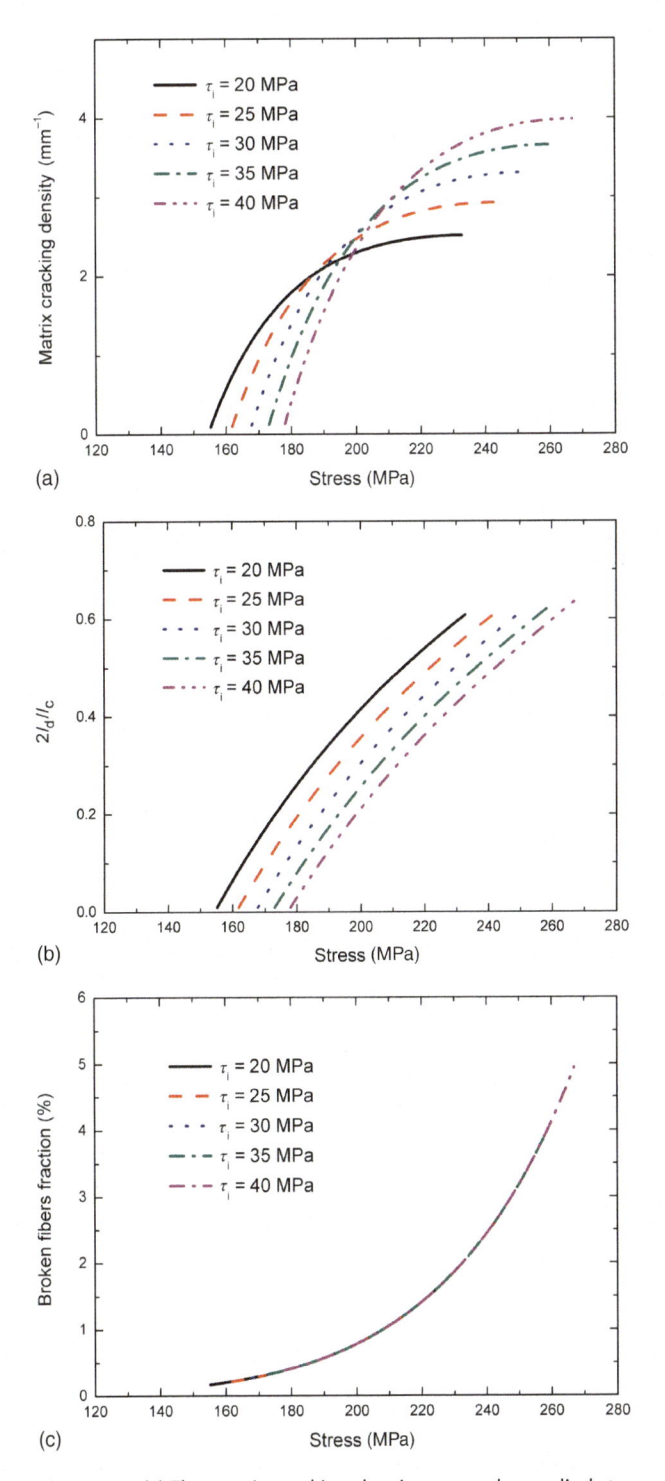

Figure 2.3 (a) The matrix cracking density versus the applied stress curves; (b) the interface debonding length versus the applied stress curves; and (c) the broken fibers fraction versus the applied stress curves for different interface shear stress.

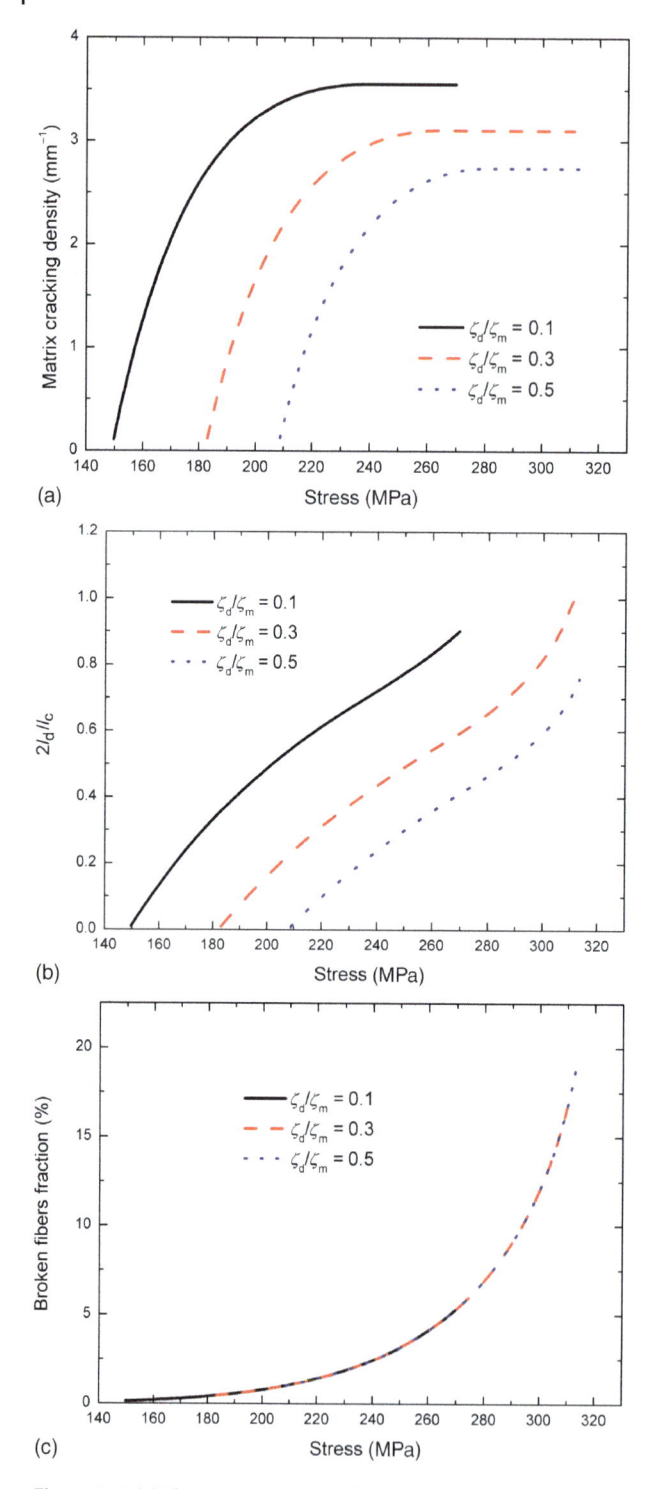

Figure 2.4 (a) The matrix cracking density versus the applied stress curves; (b) the interface debonding length versus the applied stress curves; and (c) the broken fibers fraction versus the applied stress curves for different interface debonded energy.

$\lambda = 0.13$/mm at the first matrix cracking stress of $\sigma_{mc} = 208$ MPa to $\lambda = 2.7$/mm at the saturation matrix cracking stress of $\sigma_{sat} = 280$ MPa; the interface debonding length increases from $2l_d/l_c = 0.8\%$ at the first matrix cracking stress of $\sigma_{mc} = 208$ MPa to $2l_d/l_c = 47\%$ at the saturation matrix cracking stress of $\sigma_{sat} = 280$ MPa; and the broken fibers fraction increases from $q = 1\%$ at the first matrix cracking stress of $\sigma_{mc} = 208$ MPa to $q = 7\%$ at the saturation matrix cracking stress of $\sigma_{sat} = 280$ MPa.

2.2.2.3 Effect of the Interface Properties on the Fatigue Hysteresis-Based Damage Parameters

The initial fiber/matrix interface debonding stress and the complete interface debonding stress, fatigue hysteresis dissipated energy, fatigue hysteresis modulus, and fatigue hysteresis width versus the interface shear stress and interface debonding energy curves for different fiber volume are shown in Figures 2.5 and 2.6. When the interface shear stress and the interface debonding energy increase, the initial interface debonding stress and the complete interface debonding stress increase, the fatigue hysteresis dissipated energy decreases, the fatigue hysteresis modulus increases, and the fatigue hysteresis width decreases.

When the fiber volume is $V_f = 15\%$, the initial interface debonding stress increases from $\sigma_d = 95$ MPa at the interface shear stress of $\tau_i = 20$ MPa to $\sigma_d = 98$ MPa at the interface shear stress of $\tau_i = 40$ MPa, and from $\sigma_d = 69$ MPa at the interface debonding energy of $\zeta_d/\zeta_m = 0.1$ to $\sigma_d = 201$ MPa at the interface debonding energy of $\zeta_d/\zeta_m = 0.9$; the complete interface debonding stress increases from $\sigma_b = 215$ MPa at the interface shear stress of $\tau_i = 20$ MPa to $\sigma_b = 338$ MPa at the interface shear stress of $\tau_i = 40$ MPa, and from $\sigma_b = 250$ MPa at the interface debonding energy of $\zeta_d/\zeta_m = 0.1$ to $\sigma_b = 381$ MPa at the interface debonding energy of $\zeta_d/\zeta_m = 0.9$; when the fatigue peak stress is $\sigma_{max} = 150$ MPa, the fatigue hysteresis dissipated energy decreases from $U_e = 29.1$ kJ/m^3 at the interface shear stress of $\tau_i = 20$ MPa to $U_e = 15.4$ kJ/m^3 at the interface shear stress of $\tau_i = 40$ MPa; the fatigue hysteresis modulus increases from $E = 143$ GPa at the interface shear stress of $\tau_i = 20$ MPa to $E = 202$ GPa at the interface shear stress of $\tau_i = 40$ MPa; and the fatigue hysteresis width decreases from $\Delta\varepsilon = 0.03\%$ at the interface shear stress of $\tau_i = 20$ MPa to $\Delta\varepsilon = 0.015\%$ at the interface shear stress of $\tau_i = 40$ MPa; when the fatigue peak stress is $\sigma_{max} = 250$ MPa, the fatigue hysteresis dissipated energy decreases from $U_e = 103.8$ kJ/m^3 at the interface debonding energy of $\zeta_d/\zeta_m = 0.3$ to $U_e = 49.3$ kJ/m^3 at the interface debonding energy of $\zeta_d/\zeta_m = 0.9$; the fatigue hysteresis modulus increases from $E = 130$ GPa at the interface debonding energy of $\zeta_d/\zeta_m = 0.3$ to $E = 155$ GPa at the interface debonding energy of $\zeta_d/\zeta_m = 0.9$; and the fatigue hysteresis width decreases from $\Delta\varepsilon = 0.06\%$ at the interface debonding energy of $\zeta_d/\zeta_m = 0.3$ to $\Delta\varepsilon = 0.028\%$ at the interface debonding energy of $\zeta_d/\zeta_m = 0.9$.

When the fiber volume is $V_f = 20\%$, the initial interface debonding stress increases from $\sigma_d = 131$ MPa at the interface shear stress of $\tau_i = 20$ MPa to $\sigma_d = 135$ MPa at the interface shear stress of $\tau_i = 40$ MPa, and from $\sigma_d = 95$ MPa at the interface debonding energy of $\zeta_d/\zeta_m = 0.1$ to $\sigma_d = 277$ MPa at the interface debonding energy of $\zeta_d/\zeta_m = 0.9$; the complete interface debonding stress increases from $\sigma_b = 303$ MPa at the interface shear stress of $\tau_i = 20$ MPa to

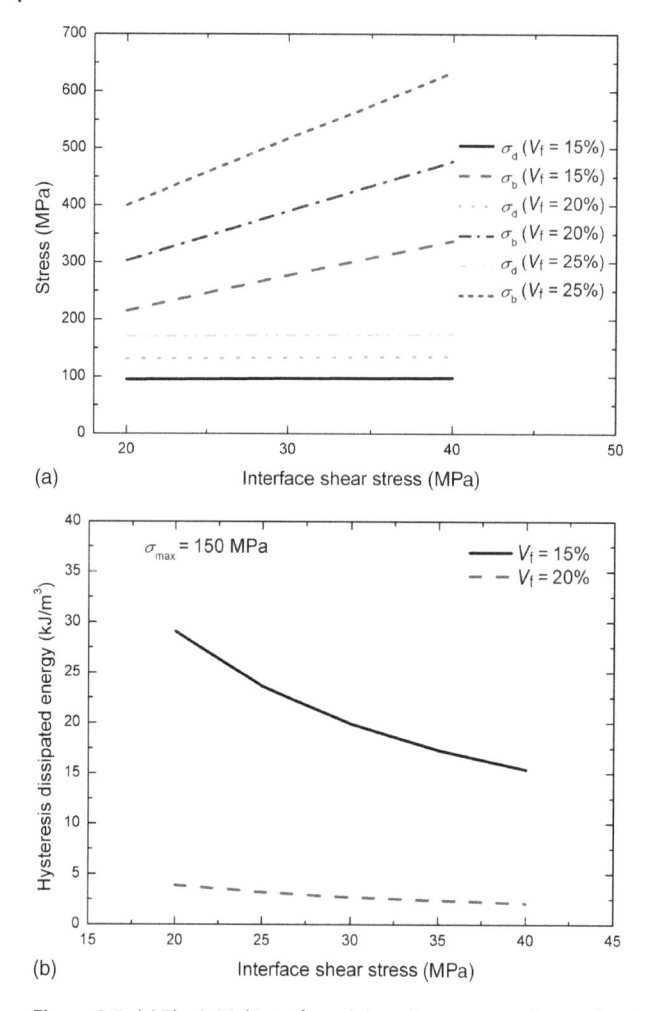

Figure 2.5 (a) The initial interface debonding stress and complete interface debonding stress versus the interface shear stress curves; (b) the fatigue hysteresis dissipated energy versus the interface shear stress curves; (c) the fatigue hysteresis modulus versus the interface shear stress curves; and (d) the fatigue hysteresis width versus the interface shear stress curves for different fiber volume.

$\sigma_b = 477$ MPa at the interface shear stress of $\tau_i = 40$ MPa, and from $\sigma_b = 352$ MPa at the interface debonding energy of $\zeta_d/\zeta_m = 0.1$ to $\sigma_b = 534$ MPa at the interface debonding energy of $\zeta_d/\zeta_m = 0.9$; when the fatigue peak stress is $\sigma_{max} = 150$ MPa, the fatigue hysteresis dissipated energy decreases from $U_e = 3.9$ kJ/m^3 at the interface shear stress of $\tau_i = 20$ MPa to $U_e = 2.1$ kJ/m^3 at the interface shear stress of $\tau_i = 40$ MPa; the fatigue hysteresis modulus increases from $E = 251$ GPa at the interface shear stress of $\tau_i = 20$ MPa to $E = 289$ GPa at the interface shear stress of $\tau_i = 40$ MPa; and the fatigue hysteresis width decreases from $\Delta\varepsilon = 0.003\%$ at the interface shear stress of $\tau_i = 20$ MPa to $\Delta\varepsilon = 0.0017\%$ at the interface shear stress of $\tau_i = 40$ MPa; when the fatigue peak stress is $\sigma_{max} = 300$ MPa,

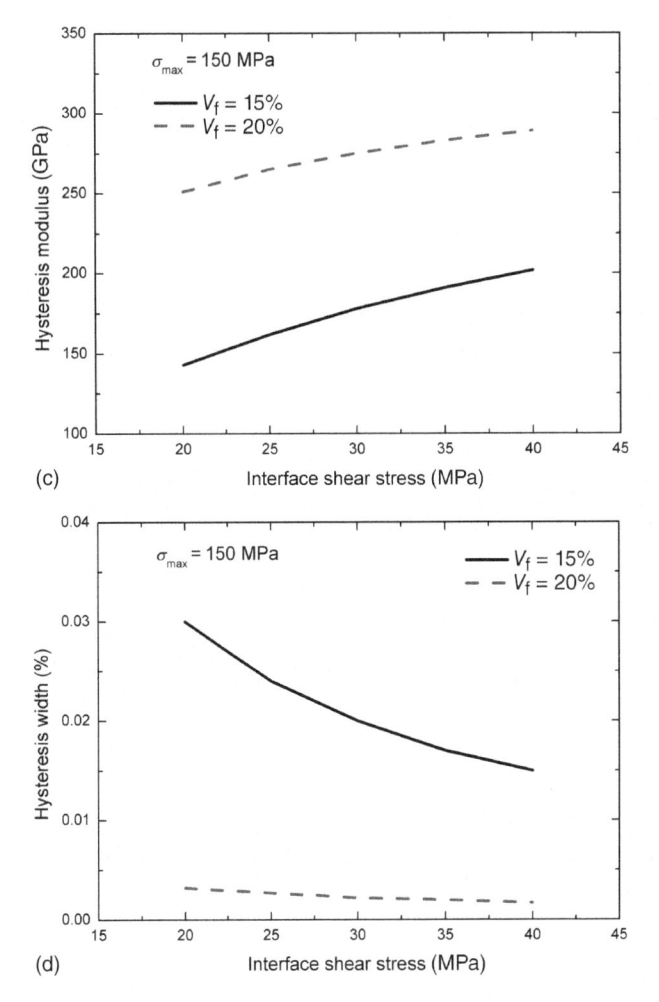

Figure 2.5 (Continued)

the fatigue hysteresis dissipated energy decreases from $U_e = 86.7\,\text{kJ/m}^3$ at the interface debonding energy of $\zeta_d/\zeta_m = 0.3$ to $U_e = 11.5\,\text{kJ/m}^3$ at the interface debonding energy of $\zeta_d/\zeta_m = 0.9$; the fatigue hysteresis modulus increases from $E = 177\,\text{GPa}$ at the interface debonding energy of $\zeta_d/\zeta_m = 0.3$ to $E = 251\,\text{GPa}$ at the interface debonding energy of $\zeta_d/\zeta_m = 0.9$; and the fatigue hysteresis width decreases from $\Delta\varepsilon = 0.043\%$ at the interface debonding energy of $\zeta_d/\zeta_m = 0.3$ to $\Delta\varepsilon = 0.004\%$ at the interface debonding energy of $\zeta_d/\zeta_m = 0.9$.

2.2.3 Experimental Comparisons

Morscher et al. (2007), Morscher and Baker (2014), and Han and Morscher (2015) performed the experimental investigations on the first matrix cracking stress, matrix cracking evolution, and fatigue hysteresis loops of 2D SiC/SiC composites. The first matrix cracking stress, matrix cracking density, and fatigue hysteresis

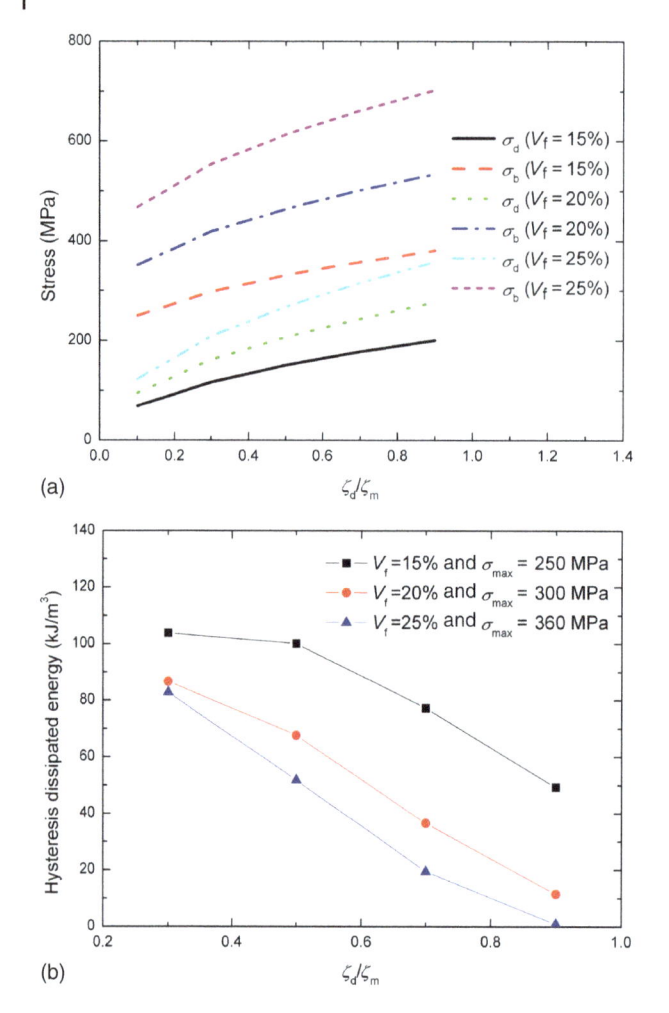

Figure 2.6 (a) The initial interface debonding stress and complete interface debonding stress versus the interface debonding energy curves; (b) the fatigue hysteresis dissipated energy versus the interface debonding energy curves; (c) the fatigue hysteresis modulus versus the interface debonding energy curves; and (d) the fatigue hysteresis width versus the interface debonding energy curves for different fiber volume.

loops of 2D Hi-Nicalon™, Sylramic™, and Tyranno™ SiC/SiC composites are predicted. The material properties of SiC/SiC composites are listed in Table 2.1.

2.2.3.1 First Matrix Cracking Stress

The experimental and predicted first matrix cracking stress versus the fiber volume curves of 2D Hi-Nicalon SiC/SiC composite for different fiber/matrix interface shear stress curves are shown in Figure 2.7. The interface shear stress is in the range of $\tau_i = 10–40$ MPa at the interface debonding energy of $\zeta_d/\zeta_m = 0.2$. The first matrix cracking stress increases with the fiber volume and the interface shear stress; the interface debonding length decreases with the fiber volume and

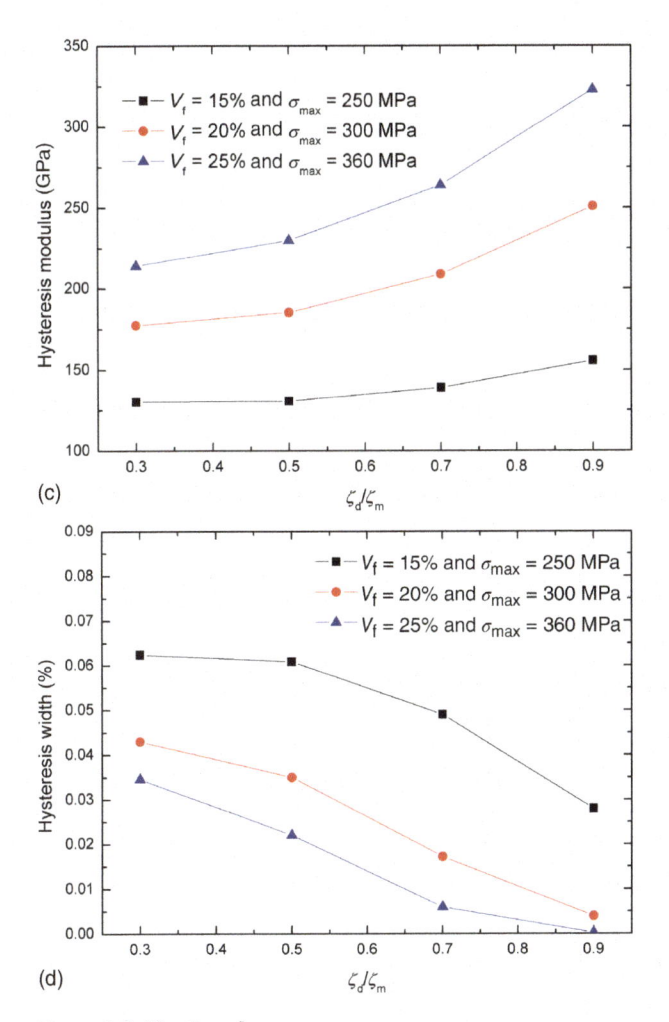

Figure 2.6 (Continued)

increases with the interface shear stress; and the broken fibers fraction decreases with the fiber volume and increases the interface shear stress.

The experimental and predicted first matrix cracking stress versus the fiber volume curves of 2D Sylramic SiC/SiC composite for different interface shear stress curves are shown in Figure 2.8. The interface shear stress is in the range of $\tau_i = 5\text{--}25$ MPa at the interface debonding energy of $\zeta_d/\zeta_m = 0.1$. The first matrix cracking stress increases with the fiber volume and the interface shear stress; the interface debonding length decreases with the fiber volume and the interface shear stress; and the broken fibers fraction decreases the fiber volume and increases with the interface shear stress.

2.2.3.2 Matrix Cracking Density

The experimental and predicted matrix cracking density versus the applied stress curves for different interface shear stress of 2D Hi-Nicalon SiC/SiC composite

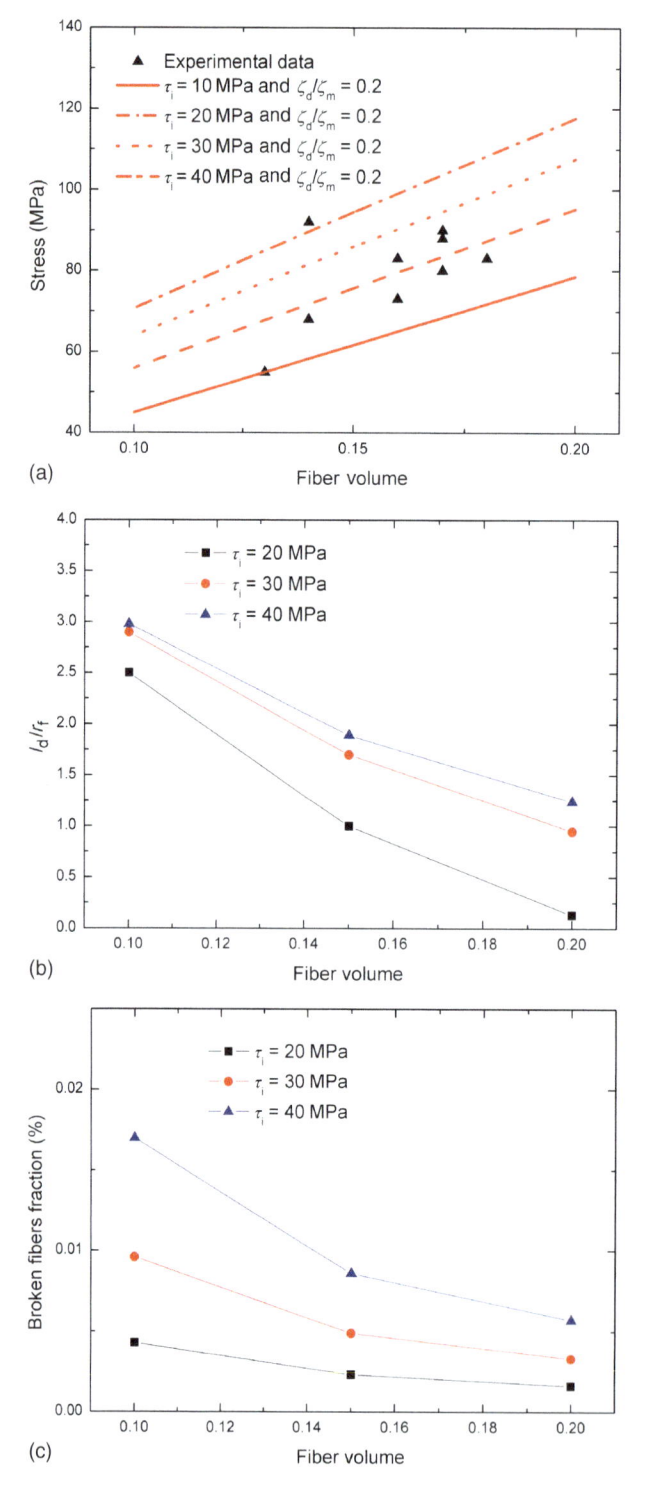

Figure 2.7 (a) The experimental and predicted first matrix cracking stress versus the fiber volume curves; (b) the interface debonding length versus the fiber volume curves; and (c) the broken fibers fraction versus the fiber volume curves of 2D Hi-Nicalon™ SiC/SiC composite.

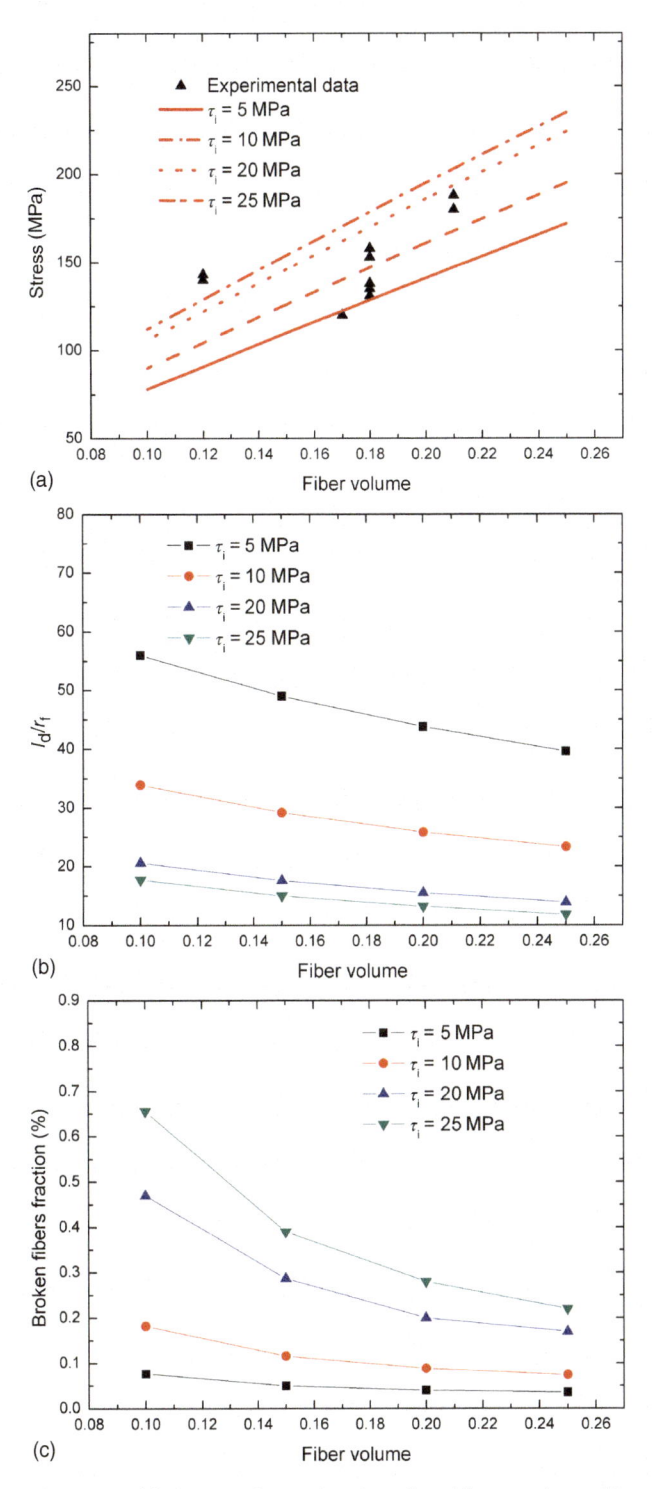

Figure 2.8 (a) The experimental and predicted first matrix cracking stress versus the fiber volume curves; (b) the interface debonding length versus the fiber volume curves; and (c) the broken fibers fraction versus the fiber volume curves of 2D Sylramic™ SiC/SiC composite.

Table 2.1 Material properties of 2D SiC/SiC composites.

Items	Hi-Nicalon™ SiC/SiC	Sylramic™ SiC/SiC	Tyranno™ SiC/SiC
r_f (μm)	7	5	5.5
V_f (%)	14–18	12–21	29
E_f (GPa)	270	310	170
E_m (GPa)	350	350	350
α_f (10^{-6}/°C)	3.5	5.4	4
α_m (10^{-6}/°C)	4.6	4.6	4.6
σ_c (GPa)	3	2.6	1.9
m_f	5	5	5

are shown in Figure 2.9. When the interface shear stress is low, the first matrix cracking stress, matrix cracking saturation stress, and saturation matrix cracking density are low. For the initial stage of matrix cracking evolution, the predicted result using low interface shear stress of $\tau_i = 20$ MPa agreed with experimental data; however, for the stage of matrix cracking evolution at high stress, the predicted results using high interface shear stress of $\tau_i = 50$ MPa agreed with experimental data. During matrix cracking evolution, the fiber/matrix interface debonding length and broken fibers fraction increase.

The experimental and predicted matrix cracking density versus the applied stress curves for different interface shear stress of 2D Sylramic SiC/SiC composite are shown in Figure 2.10. For the initial stage of matrix cracking evolution, the predicted result using the low interface shear stress of $\tau_i = 10$ and 20 MPa agreed with experimental data; however, for the stage of matrix cracking evolution at high stress, the predicted results using the high interface shear stress of $\tau_i = 50$ MPa agreed with experimental data. Under tensile loading, the interface debonding length increases after saturation of matrix cracking, and the fiber failure occurs at first matrix cracking stress.

2.2.3.3 Fatigue Hysteresis-Based Damage Parameters

The experimental and predicted hysteresis loops of 2D Sylramic SiC/SiC composite under the fatigue peak stresses of $\sigma_{max} = 200$, 240, and 275 MPa for different interface properties are shown in Figure 2.11. When the fatigue peak stress is $\sigma_{max} = 200$, 240, and 275 MPa, the predicted fatigue hysteresis loops for the interface shear stress of $\tau_i = 30$, 40, and 50 MPa and the interface debonding energy of $\zeta_d/\zeta_m = 0.1$, 0.2, and 0.3 are shown in Figure 2.11. The predicted results using the interface shear stress of $\tau_i = 50$ MPa and $\zeta_d/\zeta_m = 0.3$ agreed with the experimental hysteresis loops.

The experimental and predicted hysteresis loops of 2D Tyranno SiC/SiC composite under the fatigue peak stress of $\sigma_{max} = 120$ and 145 MPa for different interface properties are shown in Figure 2.12. When the fatigue peak stress is $\sigma_{max} = 120$ and 145 MPa, the predicted fatigue hysteresis loops for the interface shear stress of $\tau_i = 20$, 30, and 40 MPa and the interface debonding energy of

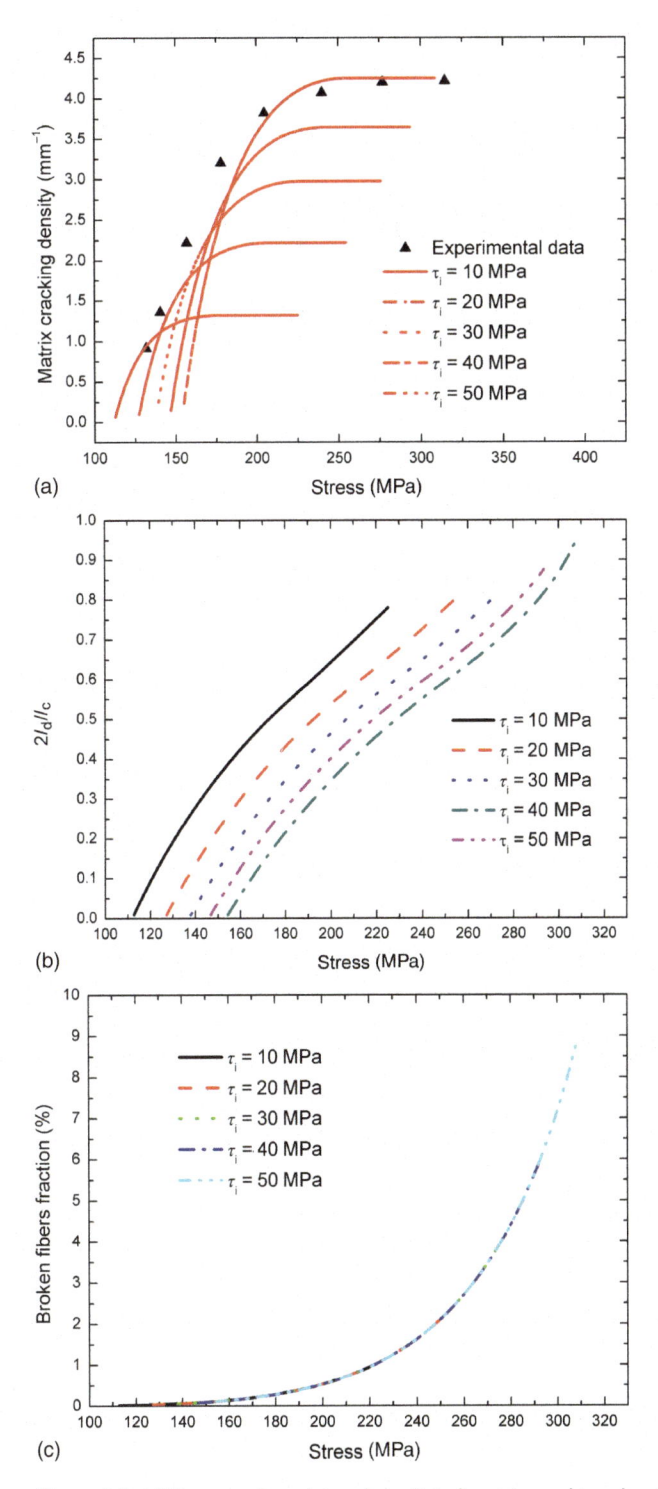

Figure 2.9 (a) The experimental and predicted matrix cracking density versus the applied stress curves; (b) the interface debonding length versus the applied stress curves; and (c) the broken fibers fraction versus the applied stress curves of 2D Hi-Nicalon™ SiC/SiC composite.

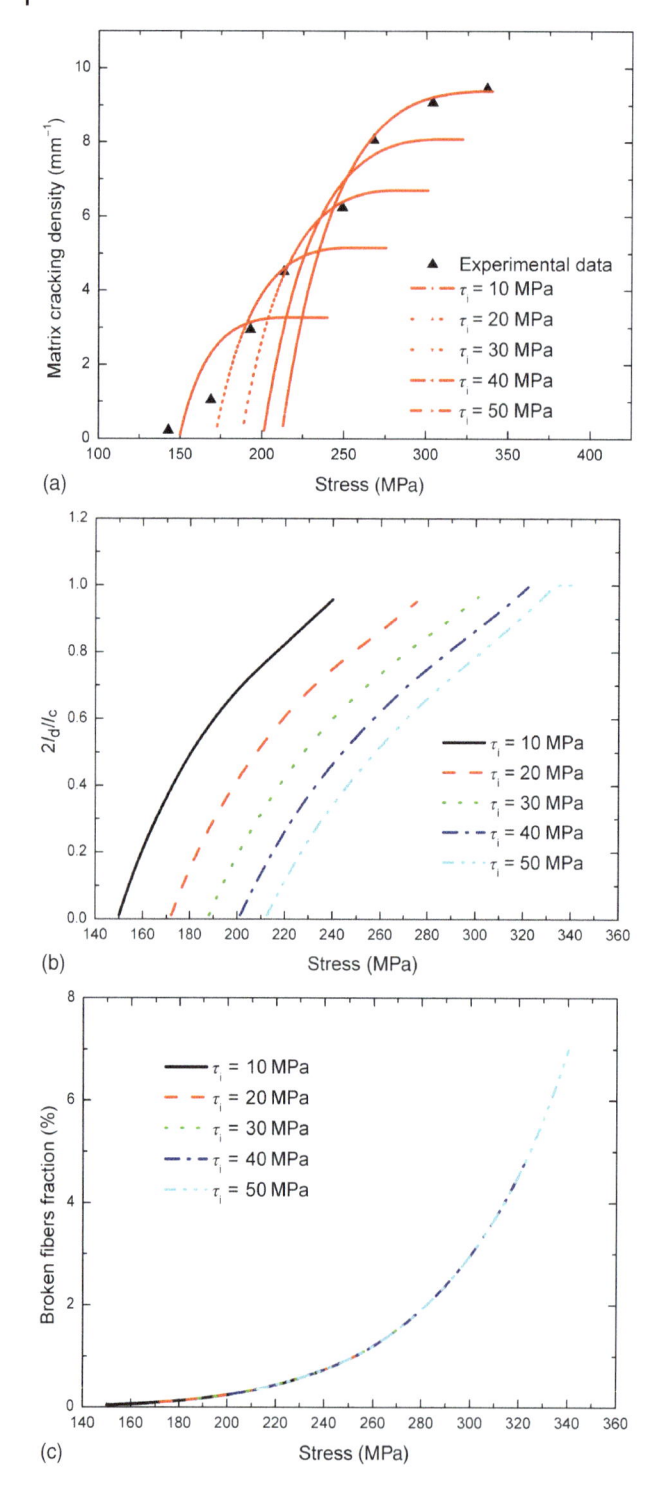

Figure 2.10 (a) The experimental and predicted matrix cracking density versus the applied stress curves; (b) the interface debonding length versus the applied stress curves; and (c) the broken fibers fraction versus the applied stress curves of 2D Sylramic™ SiC/SiC composite.

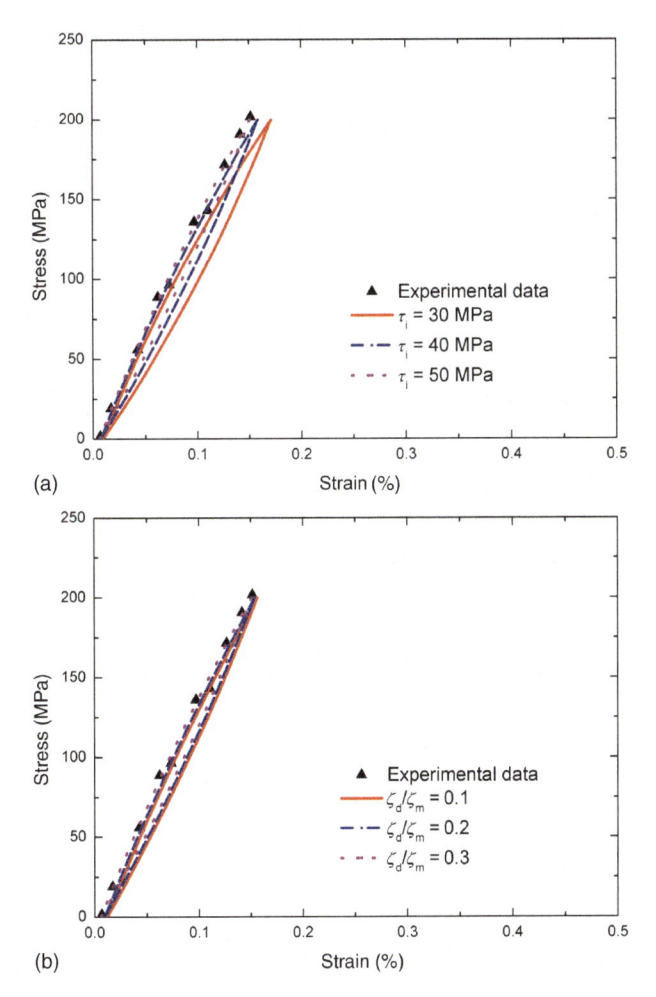

Figure 2.11 The experimental and predicted hysteresis loops of 2D Sylramic™ SiC/SiC composite under (a) the fatigue peak stress of $\sigma_{max} = 200$ MPa for different interface shear stress; (b) the fatigue peak stress of $\sigma_{max} = 200$ MPa for different interface debonding energy; (c) the fatigue peak stress of $\sigma_{max} = 240$ MPa for different interface shear stress; (d) the fatigue peak stress of $\sigma_{max} = 240$ MPa for different interface debonding energy; (e) the fatigue peak stress of $\sigma_{max} = 275$ MPa for different interface shear stress; and (f) the fatigue peak stress of $\sigma_{max} = 275$ MPa for different interface debonding energy.

$\zeta_d/\zeta_m = 0.1$, 0.2, and 0.3 are shown in Figure 2.12. The predicted results using the interface shear stress of $\tau_i = 30$ MPa and $\zeta_d/\zeta_m = 0.1$ agreed with the experimental hysteresis loops.

The fatigue hysteresis loops of 2D Hi-Nicalon SiC/SiC composite under the fatigue peak stress of $\sigma_{max} = 140$ and 150 MPa are shown in Figure 2.13. When the fatigue peak stress is $\sigma_{max} = 140$ MPa, the experimental and predicted fatigue hysteresis loops using the interface shear stress of $\tau_i = 50$, 60, 70, and 80 MPa are shown in Figure 2.13a, in which the predicted fatigue hysteresis loops with $\tau_i = 50$ MPa agreed with experimental data; when the fatigue peak stress is $\sigma_{max} = 150$ MPa, the experimental and predicted fatigue hysteresis loops using

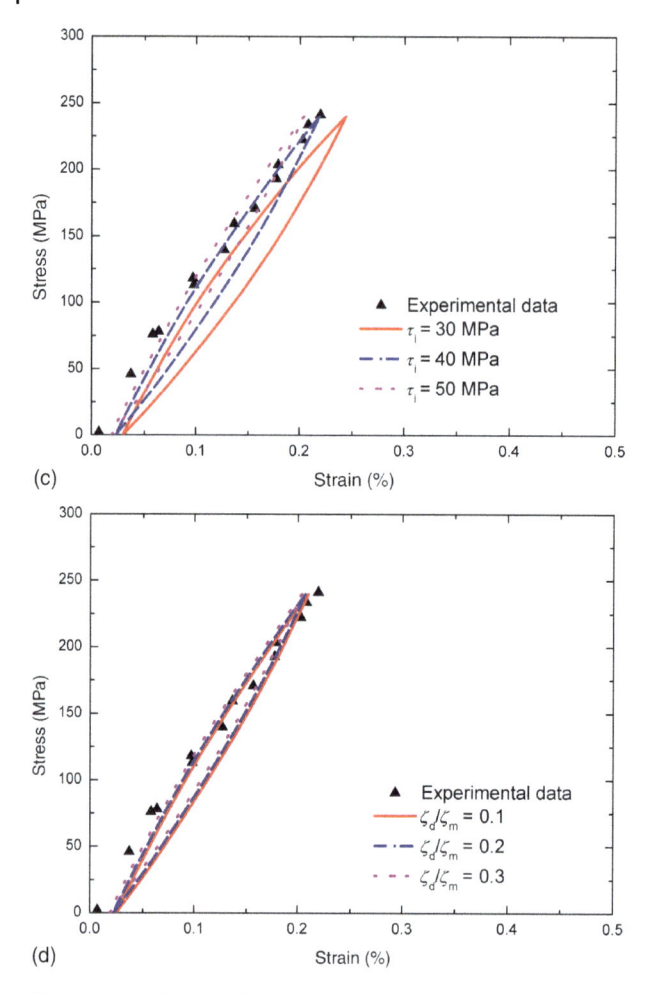

Figure 2.11 (Continued)

the interface shear stress of τ_i = 30, 40, and 50 MPa are shown in Figure 2.13b, in which the predicted fatigue hysteresis loops with τ_i = 40 MPa agreed with experimental data. Under cyclic fatigue loading, the interface wear leads to the degradation of the interface shear stress.

The fatigue hysteresis loops of 2D Tyranno SiC/SiC composite under the fatigue peak stress of σ_{max} = 230 and 240 MPa are shown in Figure 2.14. When the fatigue peak stress is σ_{max} = 230 MPa, the experimental and predicted fatigue hysteresis loops using the interface shear stress of τ_i = 40, 50, and 60 MPa are shown in Figure 2.14a, in which the predicted fatigue hysteresis loops with τ_i = 50 MPa agreed with experimental data; when the fatigue peak stress is σ_{max} = 240 MPa, the experimental and predicted fatigue hysteresis loops using the interface shear stress of τ_i = 20, 30, and 40 MPa are shown in Figure 2.14b, in which the predicted hysteresis loops with τ_i = 30 MPa agreed with experimental data. Under repeated loading/unloading, the interface shear stress decreases due to the interface wear.

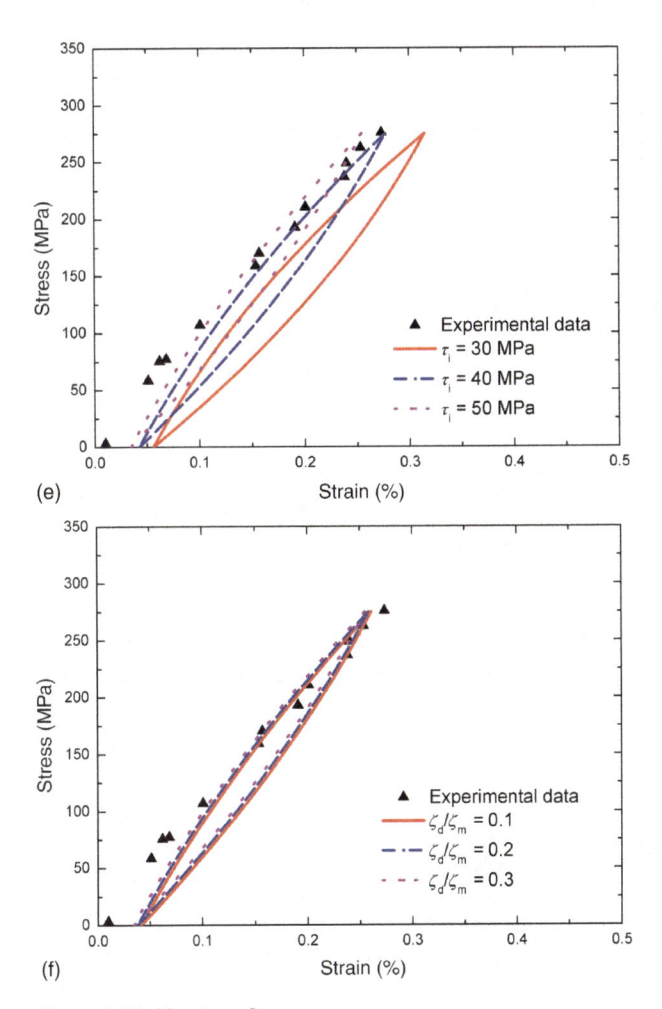

Figure 2.11 (Continued)

2.3 Effect of Pre-exposure on Tensile Damage and Fracture of Ceramic-Matrix Composites

In this section, the tensile damage and fracture process of fiber-reinforced CMCs under the effect of pre-exposure at elevated temperatures are investigated. The damage mechanisms of interface oxidation and fiber failure are considered in the stress analysis, matrix multicracking, interface debonding, and fiber failure. Combining the stress analysis and damage models, the tensile stress–strain curves of fiber-reinforced CMCs for different damage stages can be obtained. The effects of pre-exposure temperature and time, interface shear stress, fiber strength, and fiber Weibull modulus on tensile damage and fracture processes are analyzed. The experimental tensile damage and fracture process of fiber-reinforced CMCs with different fiber preforms are predicted for different pre-exposure temperature and time.

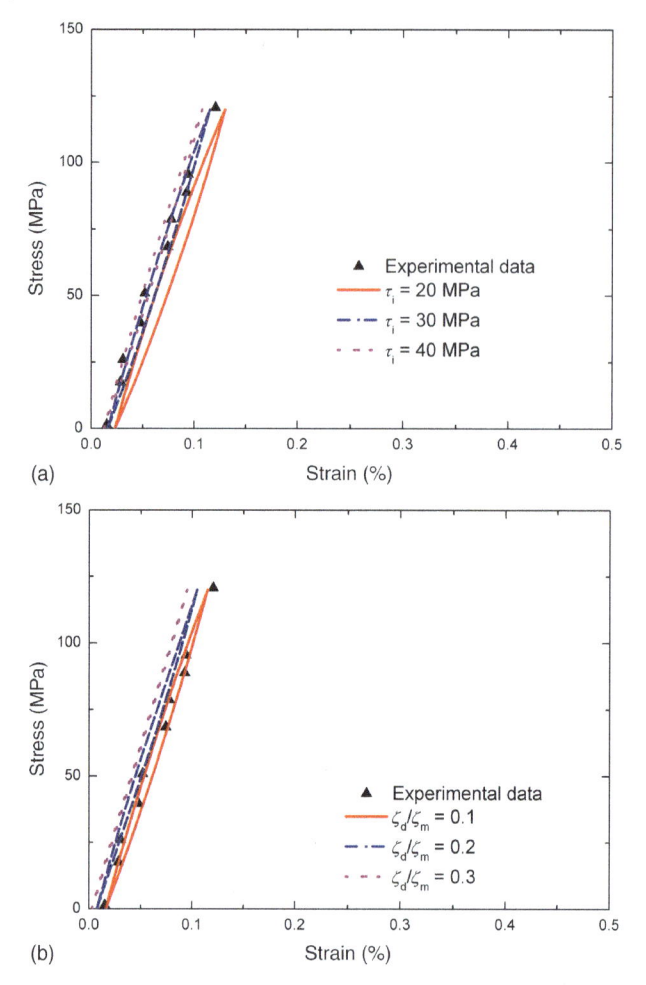

Figure 2.12 The experimental and predicted fatigue hysteresis loops of 2D Tyranno™ SiC/SiC composite under (a) the fatigue peak stress of σ_{max} = 120 MPa for different interface shear stress; (b) the fatigue peak stress of σ_{max} = 120 MPa for different interface debonding energy; (c) the fatigue peak stress of σ_{max} = 145 MPa for different interface shear stress; and (d) the fatigue peak stress of σ_{max} = 145 MPa for different interface debonding energy.

2.3.1 Theoretical Analysis

As the mismatch of the axial thermal expansion coefficient between the carbon fiber and silicon carbide matrix, there are unavoidable microcracks within SiC matrix when the composite is cooled down from high fabricated temperature to ambient temperature. These processing-induced microcracks mainly existed in the surface of the material, which do not propagate through the entire thickness of the composite. However, at elevated temperature, the microcracks serve as avenues for the ingress of the environment atmosphere into the composite. The oxygen reacts with carbon layer along the fiber length at a certain rate of $d\zeta/dt$, in which ζ is the length of carbon lost in each side of the crack (Casas and

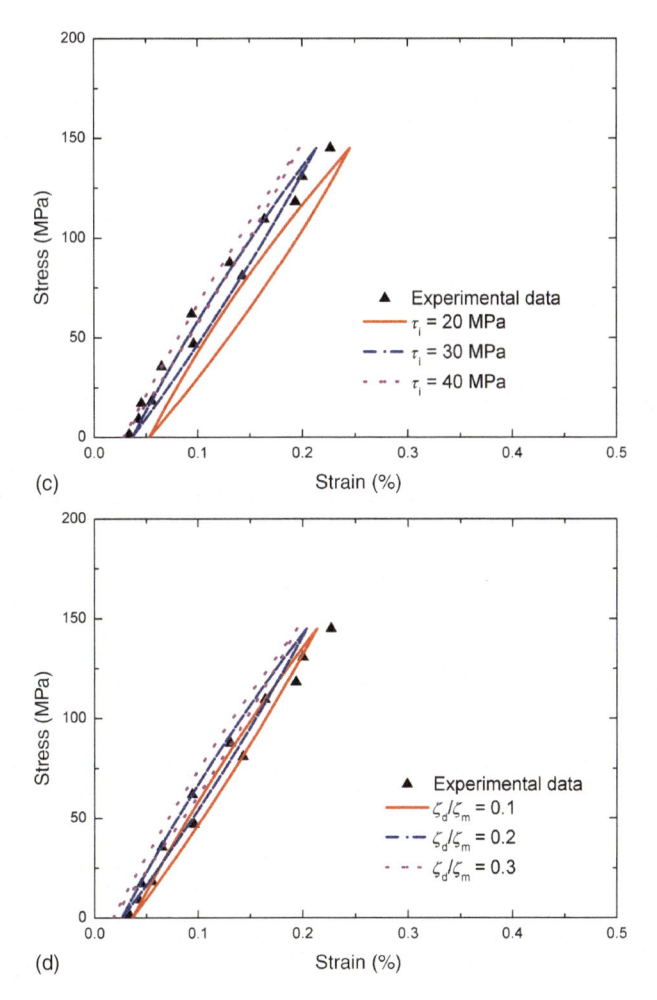

2.3 Effect of Pre-exposure on Tensile Damage and Fracture of Ceramic-Matrix Composites

Figure 2.12 (Continued)

Martinez-Esnaola 2003).

$$\zeta(t) = \varphi_1 \left[1 - \exp\left(-\frac{\varphi_2 t}{b}\right) \right] \quad (2.16)$$

where b is a delay factor considering the deceleration of reduced oxygen activity; and φ_1 and φ_2 are parameters dependent on temperature and described using the Arrhenius type laws.

$$\varphi_1 = 7.021 \times 10^{-3} \times \exp\left(\frac{8231}{T}\right) \quad (2.17)$$

$$\varphi_2 = 227.1 \times \exp\left(-\frac{17\,090}{T}\right) \quad (2.18)$$

where φ_1 is in mm and φ_2 in s^{-1}; φ_1 represents the asymptotic behavior for long times, which decreases with temperature; and the product $\varphi_1 \varphi_2$ represents the initial oxidation rate, which is an increasing function of temperature.

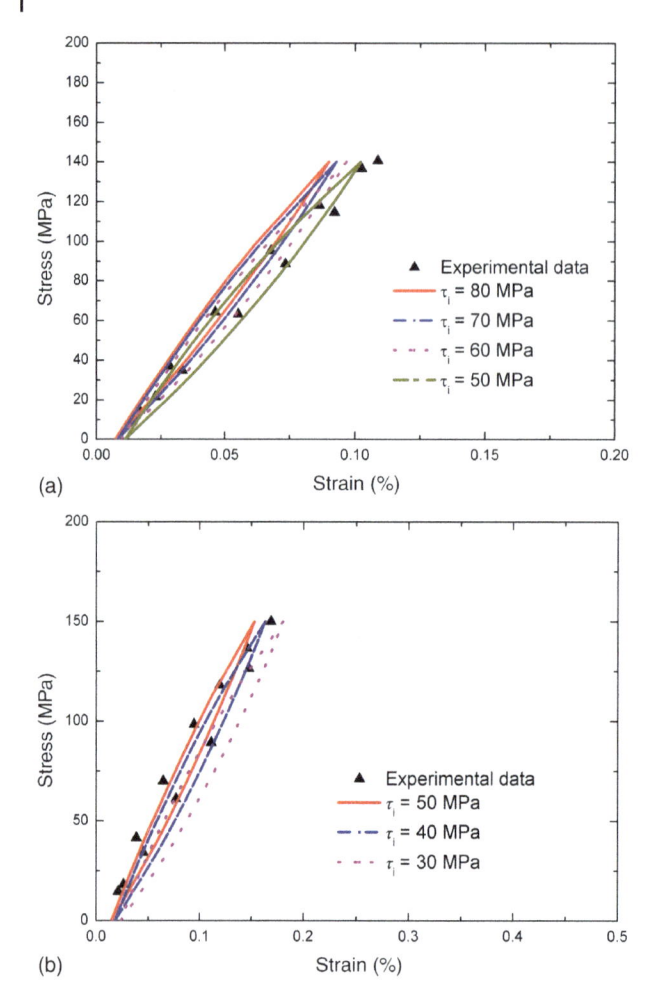

Figure 2.13 (a) The experimental and predicted fatigue hysteresis loops of 2D Hi-Nicalon™ SiC/SiC composite under the fatigue peak stress of $\sigma_{max} = 140$ MPa for $N = 13\,000$ cycles; and (b) the experimental and predicted fatigue hysteresis loops of 2D Hi-Nicalon SiC/SiC composite under the fatigue peak stress of $\sigma_{max} = 150$ MPa for $N = 27\,000$ cycles.

2.3.1.1 Stress Analysis Considering Interface Oxidation and Fiber Failure

When damage of matrix cracking and interface debonding occur in fiber-reinforced CMCs, the shear-lag model can be used to analyze the micro stress distributions in the interface oxidation region, interface slip region and interface bonded region, as shown in Figure 2.15. The distributions of the fiber and matrix axial stress distribution, and the fiber/matrix interface shear stress can be determined using the following equations:

$$\sigma_f(x) = \begin{cases} \dfrac{\sigma}{V_f} - \dfrac{2\tau_i}{r_f}x, & x \in [0, \zeta(t)] \\[6pt] \dfrac{\sigma}{V_f} - \dfrac{2\tau_f}{r_f}\zeta(t) - \dfrac{2\tau_i}{r_f}[x - \zeta(t)], & x \in [\zeta(t), l_d] \\[6pt] \sigma_{fo} + \left\{\dfrac{\sigma}{V_f} - \dfrac{2\tau_f}{r_f}\zeta(t) - \dfrac{2\tau_i}{r_f}[l_d - \zeta(t)] - \sigma_{fo}\right\}\exp\left(-\rho\dfrac{x-l_d}{r_f}\right), & x \in \left[l_d, \dfrac{l_c}{2}\right] \end{cases}$$

(2.19)

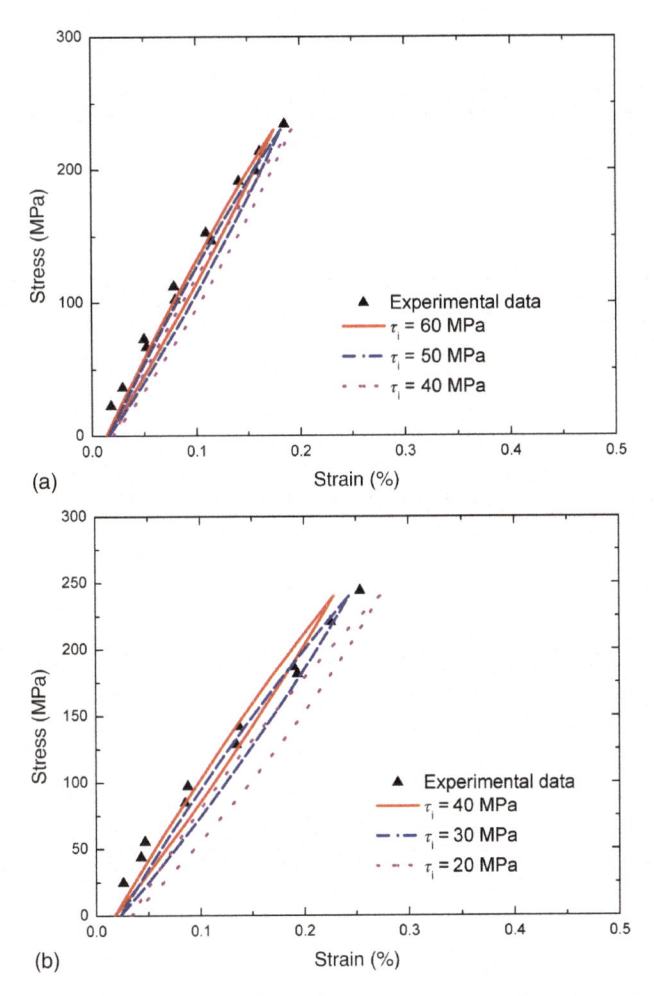

2.3 Effect of Pre-exposure on Tensile Damage and Fracture of Ceramic-Matrix Composites | 55

Figure 2.14 (a) The experimental and predicted fatigue hysteresis loops of 2D Tyranno™ SiC/SiC composite under the fatigue peak stress of $\sigma_{max} = 230$ MPa for $N = 16\,000$ cycles; and (b) the experimental and predicted fatigue hysteresis loops of 2D Tyranno SiC/SiC composite under the fatigue peak stress of $\sigma_{max} = 240$ MPa for $N = 85$ cycles.

$$\sigma_m(x) = \begin{cases} 2\dfrac{V_f}{V_m}\dfrac{\tau_f}{r_f}x, & x \in [0, \zeta(t)] \\[2mm] 2\dfrac{V_f}{V_m}\dfrac{\tau_f}{r_f}\zeta(t) + 2\dfrac{V_f}{V_m}\dfrac{\tau_i}{r_f}[x - \zeta(t)], & x \in [\zeta(t), l_d] \\[2mm] \sigma_{mo} + \left\{2\dfrac{V_f}{V_m}\dfrac{\tau_f}{r_f}\zeta(t) + 2\dfrac{V_f}{V_m}\dfrac{\tau_i}{r_f}[l_d - \zeta(t)] - \sigma_{mo}\right\} \\[2mm] \qquad \exp\left(-\rho\dfrac{x - l_d}{r_f}\right), & x \in \left[l_d, \dfrac{l_c}{2}\right] \end{cases} \qquad (2.20)$$

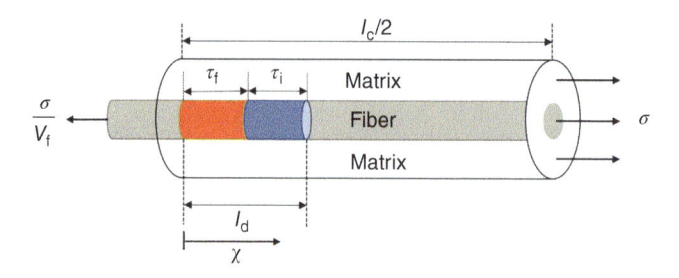

Figure 2.15 The unit cell of shear-lag model.

$$\tau_i(x) = \begin{cases} \tau_f, & x \in [0, \zeta(t)] \\ \tau_i, & x \in [\zeta(t), l_d] \\ \dfrac{\rho}{2}\left\{\dfrac{\sigma}{V_f} - \dfrac{2\tau_f}{r_f}\zeta(t) - \dfrac{2\tau_i}{r_f}[l_d - \zeta(t)] - \sigma_{fo}\right\}\exp\left(-\rho\dfrac{x - l_d}{r_f}\right), & x \in \left[l_d, \dfrac{l_c}{2}\right] \end{cases}$$

(2.21)

where V_f and V_m denote the fiber and matrix volume, respectively; τ_f and τ_i denote the interface shear stress in the interface oxidation region and interface debonded region, respectively; r_f denotes the fiber radius; l_d and l_c denote the interface debonded length and the matrix crack spacing, respectively; ρ denotes the shear-lag model parameter; and σ_{fo} and σ_{mo} denote the fiber and matrix axial stress in the interface bonded region, respectively.

$$\sigma_{fo} = \dfrac{E_f}{E_c}\sigma + E_f(\alpha_c - \alpha_f)\Delta T \tag{2.22}$$

$$\sigma_{mo} = \dfrac{E_m}{E_c}\sigma + E_m(\alpha_c - \alpha_m)\Delta T \tag{2.23}$$

where E_f, E_m, and E_c denote the fiber, matrix, and composite elastic modulus, respectively; α_f, α_m, and α_c denote the fiber, matrix, and composite thermal expansion coefficient, respectively; and ΔT denotes the temperature difference between the testing temperature and fabrication temperature.

When fiber failure occurs, the fiber axial stress in the interface debonded and bonded region can be determined using the following equation.

$$\sigma_f(x) = \begin{cases} \Phi - \dfrac{2\tau_f}{r_f}x, & x \in [0, \zeta(t)] \\ \Phi - \dfrac{2\tau_f}{r_f}\zeta(t) - \dfrac{2\tau_i}{r_f}[x - \zeta(t)], & x \in [\zeta(t), l_d] \\ \sigma_{fo} + \left\{\Phi - \dfrac{2\tau_f}{r_f}\zeta(t) - \dfrac{2\tau_i}{r_f}[l_d - \zeta(t)] - \sigma_{fo}\right\}\exp\left(-\rho\dfrac{x - l_d}{r_f}\right), \\ \qquad x \in \left[l_d, \dfrac{l_c}{2}\right] \end{cases}$$

(2.24)

2.3.1.2 Matrix Multicracking Considering Interface Oxidation

The two-parameter Weibull distribution is used to describe the tensile strength of the matrix, and the failure probability of the matrix can be determined using the following equation (Curtin 1993).

$$P_m = 1 - \exp\left\{-\left[\frac{\sigma - (\sigma_{mc} - \sigma_{th})}{(\sigma_R - \sigma_{th}) - (\sigma_{mc} - \sigma_{th})}\right]^m\right\} \quad (2.25)$$

where σ_R denotes the matrix characteristic strength; σ_{mc} denotes matrix first cracking stress; σ_{th} denotes matrix thermal residual stress; and m denotes matrix Weibull modulus.

When the applied stress increases, the matrix cracking density increases. The matrix failure probability relates with the instantaneous matrix crack space and saturation matrix crack spacing, as following:

$$P_m = l_{sat}/l_c \quad (2.26)$$

where

$$l_{sat} = \Lambda(\sigma_{mc}/\sigma_R, \sigma_{th}/\sigma_R, m)\delta_R \quad (2.27)$$

where Λ denotes the final nominal crack space, which is a pure number and depends only on the micromechanical and statistical quantities characterizing the cracking; and δ_R denotes the characteristic interface sliding length.

$$\delta_R = \frac{r_f}{2\tau_i} \frac{V_m E_m}{V_f E_c} \sigma_R + \left(1 - \frac{\tau_f}{\tau_i}\right)\zeta(t) \quad (2.28)$$

Using Eqs. (2.25)–(2.28), the instantaneous matrix crack space can be determined using the following equation.

$$l_c = r_f \frac{V_m E_m}{V_f E_c} \frac{\sigma_R}{2\tau_i} \Lambda \left\{1 - \exp\left[-\left(\frac{\sigma - (\sigma_{mc} - \sigma_{th})}{(\sigma_R - \sigma_{th}) - (\sigma_{mc} - \sigma_{th})}\right)^m\right]\right\}^{-1} \quad (2.29)$$

2.3.1.3 Interface Debonding Considering Interface Oxidation

When the matrix cracking propagates to the fiber/matrix interface, the fracture mechanics approach is used to determine the interface debonded length (Gao et al. 1988).

$$\xi_d = -\frac{F}{4\pi r_f} \frac{\partial w_f(x=0)}{\partial l_d} - \frac{1}{2}\int_0^{l_d} \tau_i \frac{\partial v(x)}{\partial l_d} dx \quad (2.30)$$

where ξ_d denotes the interface debond energy; $F(\pi r_f^2 \sigma/V_f)$ denotes the fiber stress on the matrix cracking plane; $w_f(x=0)$ denotes the fiber axial displacement at the matrix cracking plane; and $v(x)$ denotes the relative displacement between the fiber and the matrix.

$$w_f(x) = \int_x^{l_c/2} \frac{\sigma_f(x)}{E_f} dx$$

$$= \frac{\sigma}{V_f E_f}(l_d - x) - \frac{\tau_f}{r_f E_f}[2\zeta(t)l_d - \zeta^2 - x^2] - \frac{\tau_i}{r_f E_f}[l_d - \zeta(t)]^2 + \frac{\sigma_{fo}}{E_f}\left(\frac{l_c}{2} - l_d\right)$$

$$+ \frac{r_f}{\rho E_f}\left[\frac{V_m}{V_f}\sigma_{mo} - \frac{2\tau_f}{r_f}\zeta(t) - \frac{2\tau_i}{r_f}(l_d - \zeta(t))\right]\left[1 - \exp\left(-\rho\frac{l_c/2 - l_d}{r_f}\right)\right] \quad (2.31)$$

$$w_m(x) = \int_x^{l_c/2} \frac{\sigma_m(x)}{E_m} dx$$

$$= \frac{V_f \tau_f}{r_f V_m E_m}(2\zeta(t)l_d - \zeta^2(t) - x^2) + \frac{V_f \tau_i}{r_f V_m E_m}(l_d - \zeta(t))^2 + \frac{\sigma_{mo}}{E_m}\left(\frac{l_c}{2} - l_d\right)$$

$$- \frac{r_f}{\rho E_m}\left[\sigma_{mo} - 2\frac{V_f \tau_f}{r_f V_m}\zeta(t) - 2\frac{V_f \tau_i}{r_f V_m}(l_d - \zeta(t))\right]\left[1 - \exp\left(-\rho\frac{l_c/2 - l_d}{r_f}\right)\right]$$

(2.32)

The relative displacement $v(x)$ between the fiber and the matrix is described using the following equation.

$$v(x) = |w_f(x) - w_m(x)|$$

$$= \frac{\sigma}{V_f E_f}(l_d - x) - \frac{E_c \tau_f}{r_f V_m E_m E_f}(2\zeta(t)l_d - \zeta^2(t) - x^2) - \frac{E_c \tau_i}{r_f V_m E_m E_f}(l_d - \zeta(t))^2$$

$$+ \frac{r_f E_c}{\rho V_m E_m E_f}\left[\sigma_{mo} - 2\frac{\tau_f}{r_f}\zeta(t) - 2\frac{\tau_i}{r_f}(l_d - \zeta(t))\right]\left[1 - \exp\left(-\rho\frac{l_c/2 - l_d}{r_f}\right)\right]$$

(2.33)

Substituting $w_f(x=0)$ and $v(x)$ into Eq. (2.30), it leads to the following equation.

$$\frac{E_c \tau_i^2}{r_f V_m E_m E_f}(l_d - \zeta(t))^2 + \frac{E_c \tau_i^2}{\rho V_m E_m E_f}(l_d - \zeta(t)) - \frac{\tau_i \sigma}{V_f E_f}(l_d - \zeta(t))$$

$$+ \frac{2E_c \tau_f \tau_i}{r_f V_m E_m E_f}\zeta(t)(l_d - \zeta(t)) - \frac{r_f \tau_i \sigma}{2\rho V_f E_f} + \frac{E_c \tau_f^2}{r_f V_m E_m E_f}\zeta^2(t)$$

$$+ \frac{E_c \tau_f \tau_i}{\rho V_m E_m E_f}\zeta(t) - \frac{\tau_f \sigma}{V_f E_f}\zeta(t) + \frac{r_f V_m E_m \sigma^2}{4V_f^2 E_f E_c} - \xi_d = 0 \qquad (2.34)$$

Solve Eq. (2.34); the interface debonded length is determined using the following equation.

$$l_d = \left(1 - \frac{\tau_f}{\tau_i}\right)\zeta(t) + \frac{r_f}{2}\left(\frac{V_m E_m \sigma}{V_f E_c \tau_i} - \frac{1}{\rho}\right) - \sqrt{\left(\frac{r_f}{2\rho}\right)^2 + \frac{r_f V_m E_m E_f}{E_c \tau_i^2}\xi_d}$$

(2.35)

2.3.1.4 Fiber Failure Considering Interface and Fiber Oxidation

The two-parameter Weibull model is adopted to describe the fiber strength distribution, and the Global Load Sharing (GLS) criterion is used to determine the stress distributions between the intact and fracture fibers (Curtin 1991).

$$\frac{\sigma}{V_f} = \Phi(1 - q(\Phi)) + \frac{2\tau_f}{r_f}\langle L \rangle q(\Phi) \qquad (2.36)$$

where $\langle L \rangle$ denotes the average fiber pullout length; and $q(\Phi)$ denotes the fiber failure probability.

$$q(\Phi) = 1 - \exp\left[-\left(\frac{\Phi}{\sigma_c}\right)^{m_f+1}\right] \qquad (2.37)$$

where m_f denotes the fiber Weibull modulus; σ_c denotes the fiber characteristic strength of a length δ_c of fiber.

$$\sigma_c = \left(\frac{l_0 \sigma_0^{m_f} \tau_i}{r_f}\right)^{\frac{1}{m_f+1}}, \delta_c = \left[\frac{\sigma_0 r_f l_0^{m_f}}{\tau_i}\right]^{\frac{1}{m_f+1}\left(\frac{m_f}{m_f+1}\right)} \quad (2.38)$$

The time-dependent fiber strength of $\sigma_0(t)$ can be determined using the following equation (Lara-Curzio 1999).

$$\sigma_0(t) = \begin{cases} \sigma_0, & t \leq \frac{1}{k}\left(\frac{K_{IC}}{Y\sigma_0}\right)^4 \\ \frac{K_{IC}}{Y\sqrt[4]{kt}}, & t > \frac{1}{k}\left(\frac{K_{IC}}{Y\sigma_0}\right)^4 \end{cases} \quad (2.39)$$

The composite tensile strength is given by the following equation.

$$\sigma_{UTS} = V_f \sigma_c \left(\frac{2}{m_f+2}\right)^{\frac{1}{m_f+1}} \left(\frac{m_f+1}{m_f+2}\right) \quad (2.40)$$

2.3.1.5 Tensile Stress–Strain Curves Considering Effect of Pre-exposure

For fiber-reinforced CMCs without damage, the composite strain can be determined using the following equation.

$$\varepsilon_c = \sigma/E_c \quad (2.41)$$

When damage forms inside of CMCs, the composite strain can be determined using the following equation.

$$\varepsilon_c = \frac{2}{E_f l_c} \int_{l_c/2} \sigma_f(x) dx - (\alpha_c - \alpha_f)\Delta T \quad (2.42)$$

When matrix cracking and interface debonding occurs, the composite strain can be determined using the following equation.

$$\varepsilon_c = \frac{2\sigma}{V_f E_f} \frac{l_d}{l_c} + \frac{2\tau_f}{r_f E_f l_c}\zeta^2(t) - \frac{4\tau_f}{r_f E_f} \frac{l_d}{l_c}\zeta(t) - \frac{2\tau_i}{r_f E_f l_c}[l_d - \zeta(t)]^2 + \frac{2\sigma_{fo}}{E_f l_c}\left(\frac{l_c}{2} - l_d\right)$$
$$+ \frac{2r_f}{\rho E_f l_c}\left\{\frac{\sigma}{V_f} - \frac{2\tau_f}{r_f}\zeta(t) - \frac{2\tau_i}{r_f}[l_d - \zeta(t)] - \sigma_{fo}\right\}$$
$$\times \left[1 - \exp\left(-\rho\frac{l_c/2 - l_d}{r_f}\right)\right] - (\alpha_c - \alpha_f)\Delta T \quad (2.43)$$

When fiber failure occurs, the composite strain can be determined using the following equation.

$$\varepsilon_c = \frac{\Phi}{E_f} \frac{2l_d}{l_c} + \frac{2\tau_f}{r_f E_f l_c} \zeta^2(t) - \frac{4\tau_f l_d}{r_f E_f l_c} \zeta(t) - \frac{2\tau_i}{r_f E_f l_c}(l_d - \zeta(t))^2 + \frac{2\sigma_{fo}}{E_f l_c}\left(\frac{l_c}{2} - l_d\right)$$
$$+ \frac{2r_f}{\rho E_f l_c}\left\{\Phi - \frac{2\tau_f}{r_f}\zeta(t) - \frac{2\tau_i}{r_f}[l_d - \zeta(t)] - \sigma_{fo}\right\}$$
$$\times \left[1 - \exp\left(-\rho\frac{l_c/2 - l_d}{r_f}\right)\right] - (\alpha_c - \alpha_f)\Delta T \tag{2.44}$$

For the 2D, 2.5D, and 3D CMCs, an effective coefficient of fiber volume fraction along the loading direction (ECFL) is defined as

$$\Lambda = \frac{V_f^{\text{axial}}}{V_f} \tag{2.45}$$

where V_f and V_f^{axial} refer to the total fiber volume fraction in the composites and the effective fiber volume fraction in the direction of axial tensile loading.

2.3.2 Results and Discussion

The effects of pre-exposure temperature and time, interface shear stress, fiber strength, and fiber Weibull modulus on tensile damage process are analyzed. The unidirectional C/SiC composite is used to the case analysis, and the material properties are given by $V_f = 40\%$, $E_f = 230$ GPa, $E_m = 350$ GPa, $r_f = 3.5$ µm, $m = 6$, $\sigma_R = 100$ MPa, $l_{sat} = 150$ µm, $\alpha_f = 0 \times 10^{-6}$/K, $\alpha_m = 4.6 \times 10^{-6}$/K, $\Delta T = -1000$ °C, $\zeta_d = 0.1$ J/m², $\tau_i = 10$ MPa, $\tau_f = 1$ MPa, $\sigma_c = 1.6$ GPa, and $m_f = 5$.

2.3.2.1 Effect of Pre-exposure Temperature on Tensile and Damage Process

The effect of pre-exposure temperature (i.e. $T = 600$, 700, and 800 °C) on the tensile stress–strain curves, interface debonding and oxidation, and fiber failure of C/SiC composite corresponding to pre-exposure time of $t = 20$ hours is shown in Figure 2.16. When the pre-exposure temperature increases, the composite tensile strength and failure strain both decrease; the interface debonded length and the interface oxidation ratio both increase; and the fiber broken fraction increases at low applied stress level.

When the pre-exposure temperature is $T = 600$ °C, the composite tensile strength is $\sigma_{\text{UTS}} = 445$ MPa with the failure strain of $\varepsilon_f = 0.57\%$; the interface debonding length increases to $2l_d/l_c = 67.5\%$; the interface oxidation ratio decreases to $\zeta/l_d = 2.9\%$; and the fiber broken fraction increases to $q = 13.3\%$.

When the pre-exposure temperature is $T = 700$ °C, the composite tensile strength is $\sigma_{\text{UTS}} = 366$ MPa with the failure strain of $\varepsilon_f = 0.45\%$; the interface debonding length increases to $2l_d/l_c = 55\%$; the interface oxidation ratio decreases to $\zeta/l_d = 10.3\%$; and the fiber broken fraction increases to $q = 12.1\%$.

When the pre-exposure temperature is $T = 800$ °C, the composite tensile strength is $\sigma_{\text{UTS}} = 297$ MPa with the failure strain of $\varepsilon_f = 0.37\%$; the interface debonding length increases to $2l_d/l_c = 48\%$; the interface oxidation ratio decreases to $\zeta/l_d = 27.5\%$; and the fiber broken fraction increases to $q = 12\%$.

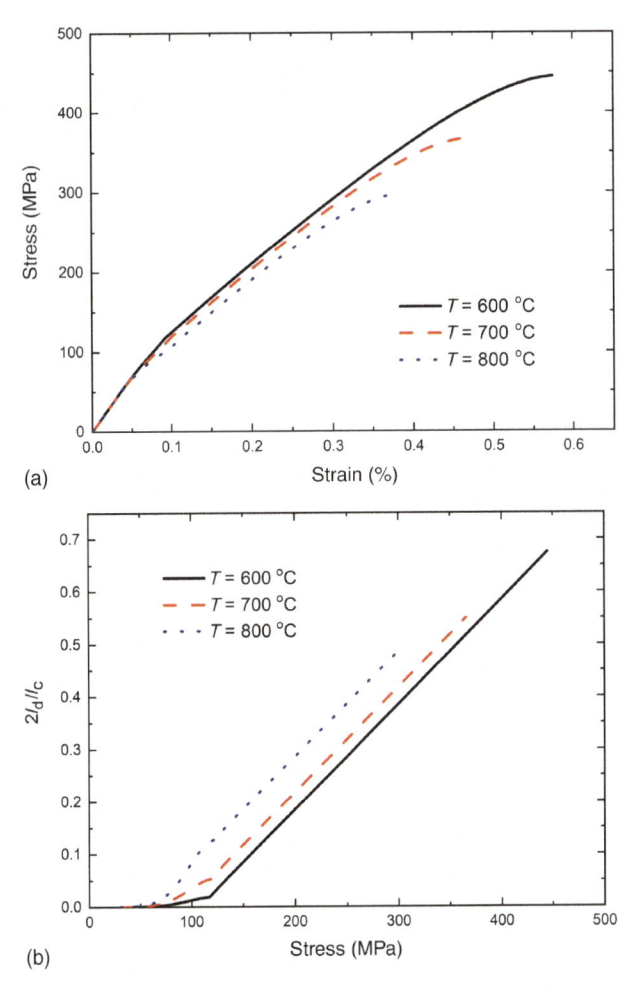

Figure 2.16 The effect of pre-exposure temperature on (a) the tensile stress–strain curves; (b) the interface debonded length of $2l_d/l_c$ versus the applied stress curves; (c) the interface oxidation ratio of ζ/l_d versus the applied stress curves; and (d) the broken fibers fraction of q versus applied stress curves of C/SiC composite.

2.3.2.2 Effect of Pre-exposure Time on Tensile and Damage Processes

The effect of pre-exposure time (i.e. $t = 10$, 20, and 30 hours) on the tensile stress–strain curves, interface debonding and oxidation, and fiber failure of C/SiC composite corresponding to pre-exposure temperature of $T = 800\,°C$ is shown in Figure 2.17. When the pre-exposure time increases, the composite tensile strength and failure strain both decrease; the interface debonded length and the interface oxidation ratio both increase; and the fiber broken fraction increases at low applied stress level.

When the pre-exposure time is $t = 10$ hours, the composite tensile strength is $\sigma_{UTS} = 354$ MPa with the failure strain of $\varepsilon_f = 0.45\%$; the interface debonding

(c) [Graph: ζ/l_d vs Stress (MPa), curves for T = 600 °C, T = 700 °C, T = 800 °C]

(d) [Graph: Broken fibers fraction (%) vs Stress (MPa), curves for T = 600 °C, T = 700 °C, T = 800 °C]

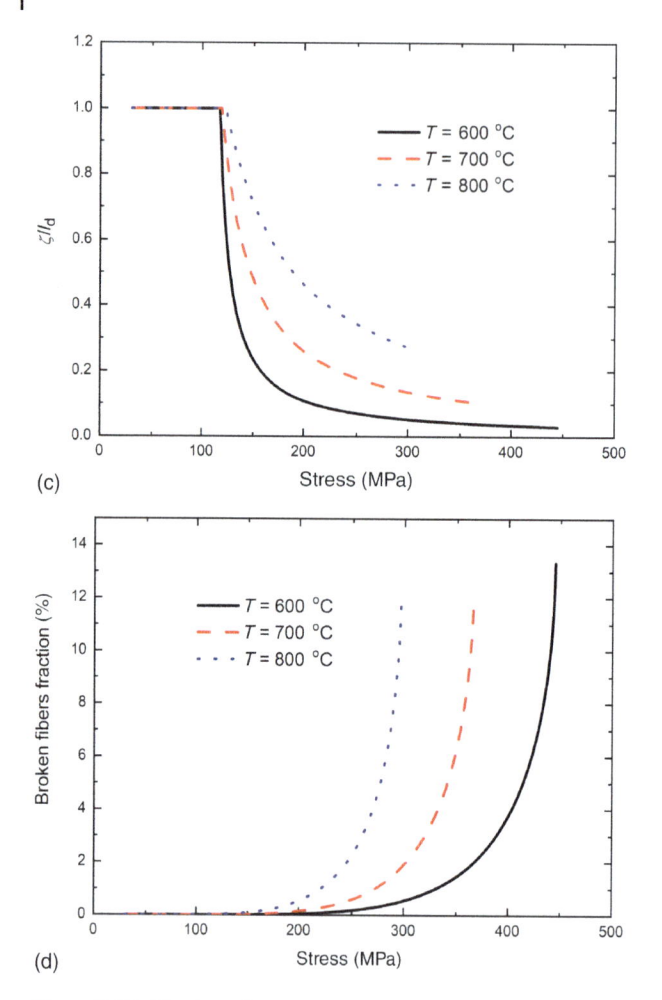

Figure 2.16 (Continued)

length increases to $2l_d/l_c = 53\%$; the interface oxidation ratio decreases to $\zeta/l_d = 12.3\%$; and the fiber broken fraction increases to $q = 13.4\%$.

When the pre-exposure time is $t = 20$ hours, the composite tensile strength is $\sigma_{UTS} = 297$ MPa with the failure strain of $\varepsilon_f = 0.37\%$; the interface debonding length increases to $2l_d/l_c = 48\%$; the interface oxidation ratio decreases to $\zeta/l_d = 27.5\%$; and the fiber broken fraction increases to $q = 12\%$.

When the pre-exposure time is $t = 30$ hours, the composite tensile strength is $\sigma_{UTS} = 269$ MPa with the failure strain of $\varepsilon_f = 0.35\%$; the interface debonding length increases to $2l_d/l_c = 48.4\%$; the interface oxidation ratio decreases to $\zeta/l_d = 41\%$; and the fiber broken fraction increases to $q = 13.5\%$.

2.3.2.3 Effect of Interface Shear Stress on Tensile and Damage Processes

The effect of interface shear stress (i.e. $\tau_i = 5$, 10, and 15 MPa) on the tensile stress–strain curves, interface debonding and oxidation, and fiber failure of

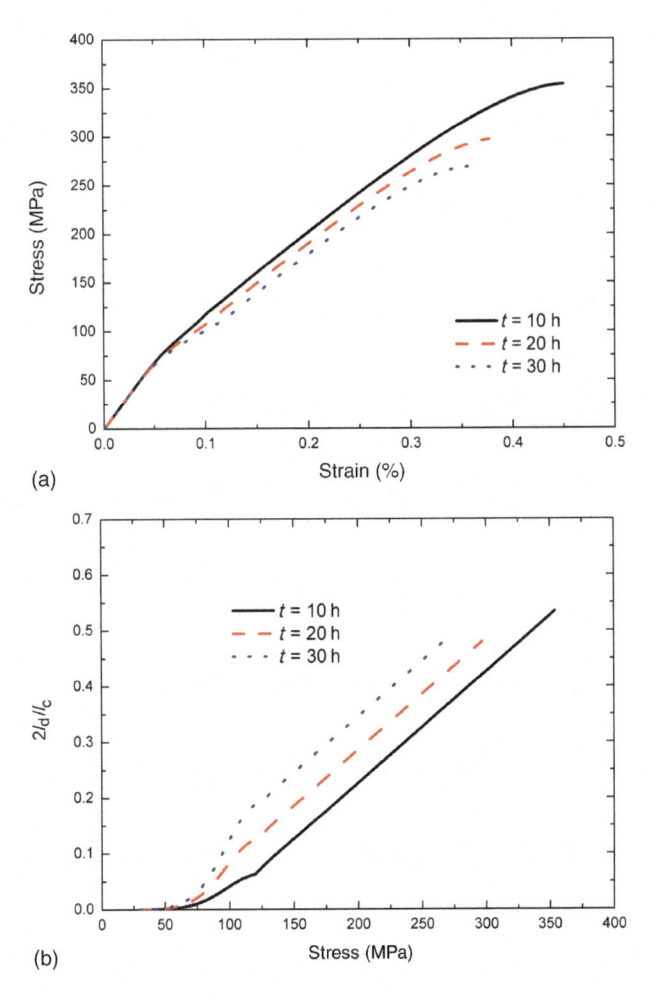

Figure 2.17 The effect of pre-exposure time on (a) the tensile stress–strain curves; (b) the interface debonded length of $2l_d/l_c$ versus the applied stress curves; (c) the interface oxidation ratio of ζ/l_d versus the applied stress curves; and (d) the broken fibers fraction versus applied stress curves of C/SiC composite.

C/SiC composite corresponding to pre-exposure temperature of $T = 800\,°C$ and pre-exposure time of $t = 20$ hours is shown in Figure 2.18. When the interface shear stress increases, the composite failure strain decreases; the interface debonded length decreases, and the interface oxidation ratio increases.

When the interface shear stress is $\tau_i = 5\,\text{MPa}$, the composite failure strain is $\varepsilon_f = 0.46\%$; the interface debonding length increases to $2l_d/l_c = 84\%$; the interface oxidation ratio decreases to $\zeta/l_d = 15.7\%$; and the fiber broken fraction increases to $q = 12\%$.

When the interface shear stress is $\tau_i = 10\,\text{MPa}$, the composite failure strain is $\varepsilon_f = 0.37\%$; the interface debonding length increases to $2l_d/l_c = 48\%$; the interface oxidation ratio decreases to $\zeta/l_d = 27\%$; and the fiber broken fraction increases to $q = 12\%$.

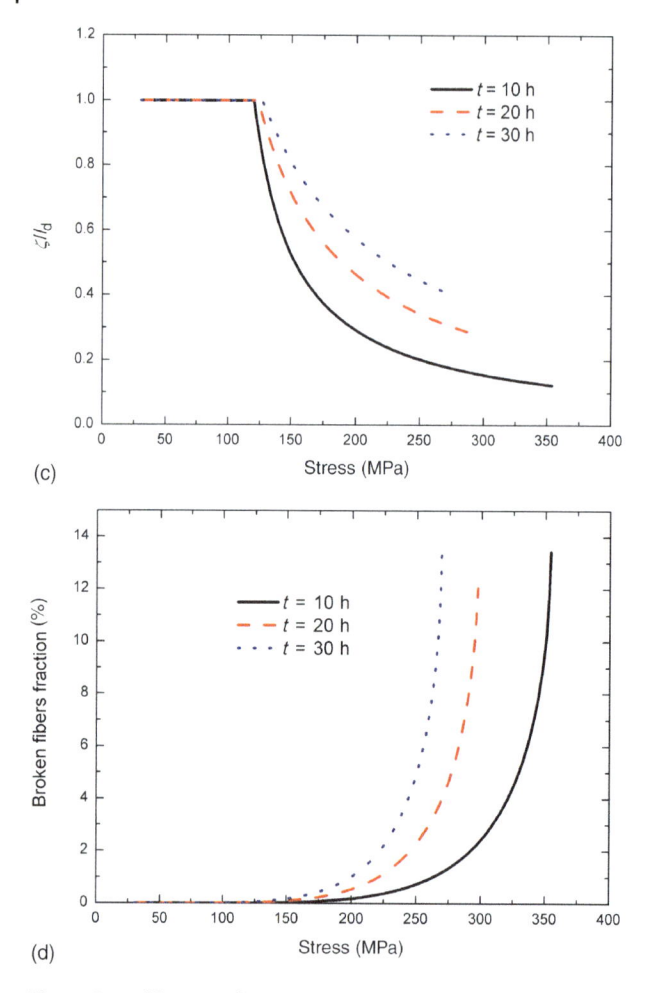

Figure 2.17 (Continued)

When the interface shear stress is $\tau_i = 15$ MPa, the composite failure strain is $\varepsilon_f = 0.35\%$; the interface debonding length increases to $2l_d/l_c = 36\%$; the interface oxidation ratio decreases to $\zeta/l_d = 36.7\%$; and the fiber broken fraction increases to $q = 12\%$.

2.3.2.4 Effect of Fiber Strength on Tensile and Damage Processes

The effect of fiber strength (i.e. $\sigma_0 = 1$ and 2 GPa) on the tensile stress–strain curves and fiber failure of C/SiC composite corresponding to pre-exposure temperature of $T = 800\,°C$ and pre-exposure time of $t = 20$ hours is shown in Figure 2.19. When the fiber strength increases, the composite tensile strength and failure strain both increase; and the fiber broken fraction decreases at low applied stress level.

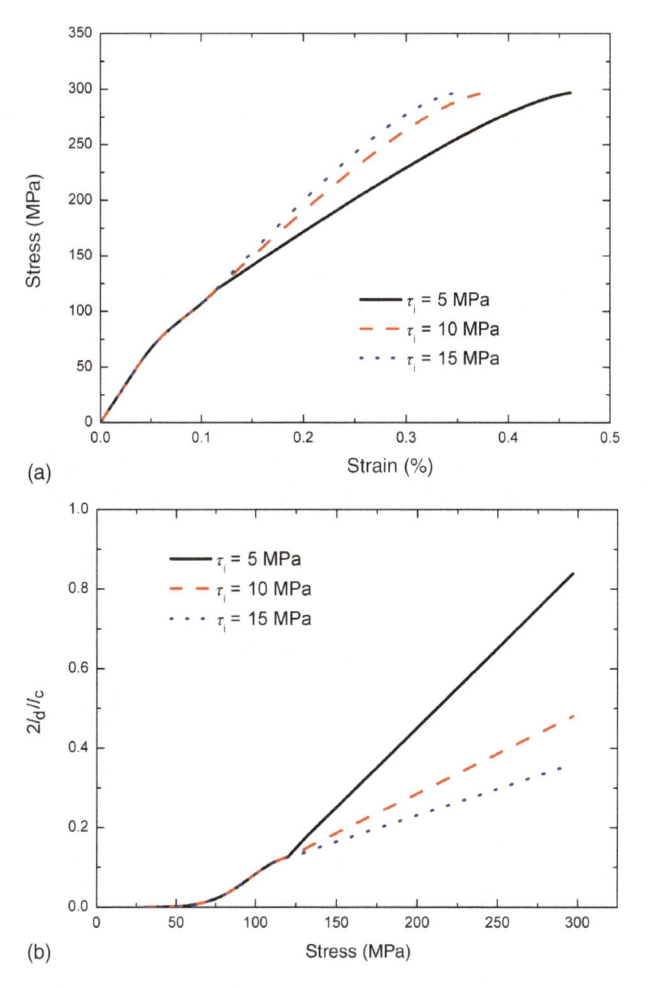

Figure 2.18 The effect of the interface shear stress on (a) the tensile stress–strain curves; (b) the interface debonded length of $2l_d/l_c$ versus the applied stress curves; (c) the interface oxidation ratio of ζ/l_d versus the applied stress curves; and (d) the broken fibers fraction versus applied stress curves of C/SiC composite.

When the fiber strength is $\sigma_0 = 1$ GPa, the composite tensile strength is $\sigma_{UTS} = 278$ MPa with the failure strain of $\varepsilon_f = 0.31\%$; and the fiber broken fraction increases to $q = 12.9\%$.

When the fiber strength is $\sigma_0 = 2$ GPa, the composite tensile strength is $\sigma_{UTS} = 354$ MPa with the failure strain of $\varepsilon_f = 0.41\%$; and the fiber broken fraction increases to $q = 13.4\%$.

2.3.2.5 Effect of Fiber Weibull Modulus on Tensile and Damage Processes

The effect of fiber Weibull modulus (i.e. $m_f = 3$ and 5) on the tensile stress–strain curves and fiber failure of C/SiC composite corresponding to pre-exposure temperature of $T = 800\,°C$ and pre-exposure time of $t = 20$ hours is shown in Figure 2.20. When the fiber Weibull modulus increases, the composite tensile

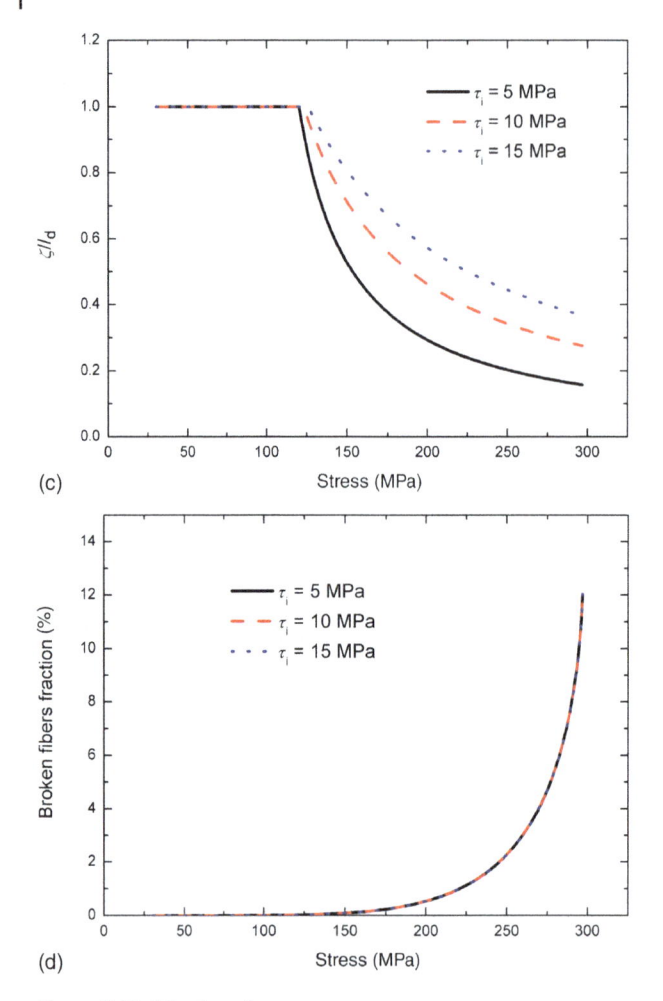

Figure 2.18 (Continued)

strength and failure strain both increase; and the fiber broken fraction decreases at low applied stress level.

When the fiber Weibull modulus is $m_f = 3$, the composite tensile strength is $\sigma_{UTS} = 272$ MPa with the failure strain of $\varepsilon_f = 0.34\%$; and the fiber broken fraction increases to $q = 18.2\%$.

When the fiber Weibull modulus is $m_f = 5$, the composite tensile strength is $\sigma_{UTS} = 297$ MPa with the failure strain of $\varepsilon_f = 0.35\%$; and the fiber broken fraction increases to $q = 12\%$.

2.3.3 Experimental Comparisons

Wang et al. (2013) investigated the tensile behavior of 1D, 2D, and 3D C/SiC composite at room temperature. Zhang et al. (2016) investigated the tensile behavior

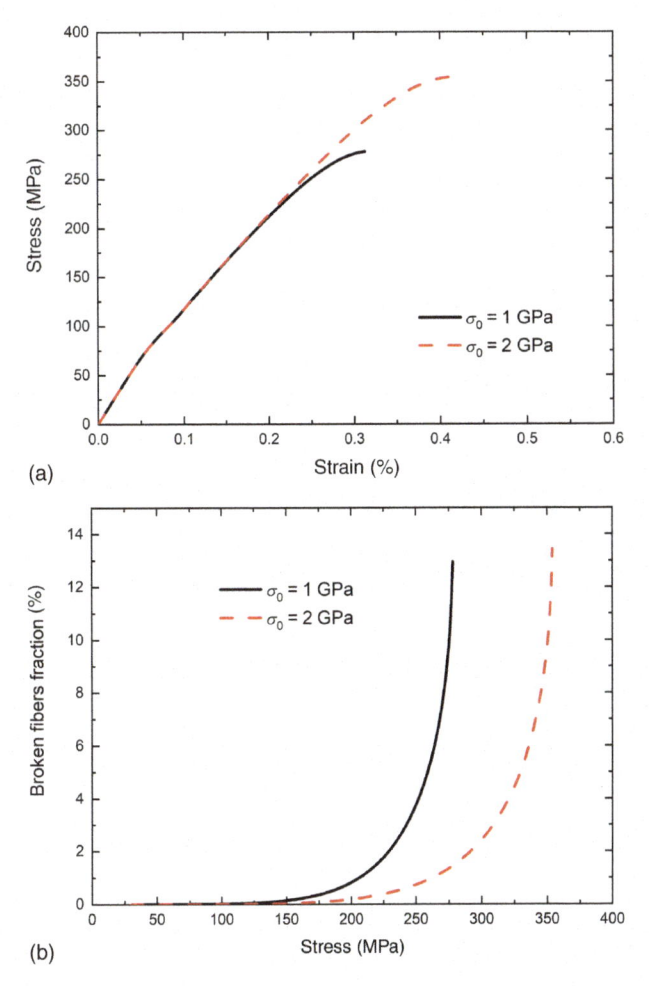

Figure 2.19 The effect of the fiber strength on (a) the tensile stress–strain curves; and (b) the broken fibers fraction versus applied stress curves of C/SiC composite.

of 2.5D C/SiC composite after exposure at elevated temperature. The material properties of 1D, 2D, 2.5D, and 3D C/SiC composites are listed in Table 2.2.

The experimental and predicted tensile stress–strain curves, interface debonded length and oxidation length, and fiber broken fraction of 1D C/SiC composite without and with pre-exposure at $T = 800\,°C$ and $t = 10, 20,$ and 30 hours are shown in Figure 2.21. With increasing of pre-exposure time, the composite tensile strength and failure strain both decrease; the interface debonded length and interface oxidation ratio increase; and the broken fibers fraction increases at low stress level. Without pre-exposure, the composite tensile strength is $\sigma_{UTS} = 333$ MPa with the failure strain of $\varepsilon_f = 0.68\%$; when the pre-exposure time is $t = 10$ hours, the composite tensile strength is $\sigma_{UTS} = 314$ MPa with the failure strain of $\varepsilon_f = 0.67\%$, the interface debonding length increases to $2l_d/l_c = 1$, the interface oxidation ratio decreases to $\zeta/l_d = 24.8\%$, and the broken fibers fraction increases to $q = 12\%$; when the pre-exposure time is $t = 20$ hours, the composite tensile strength is

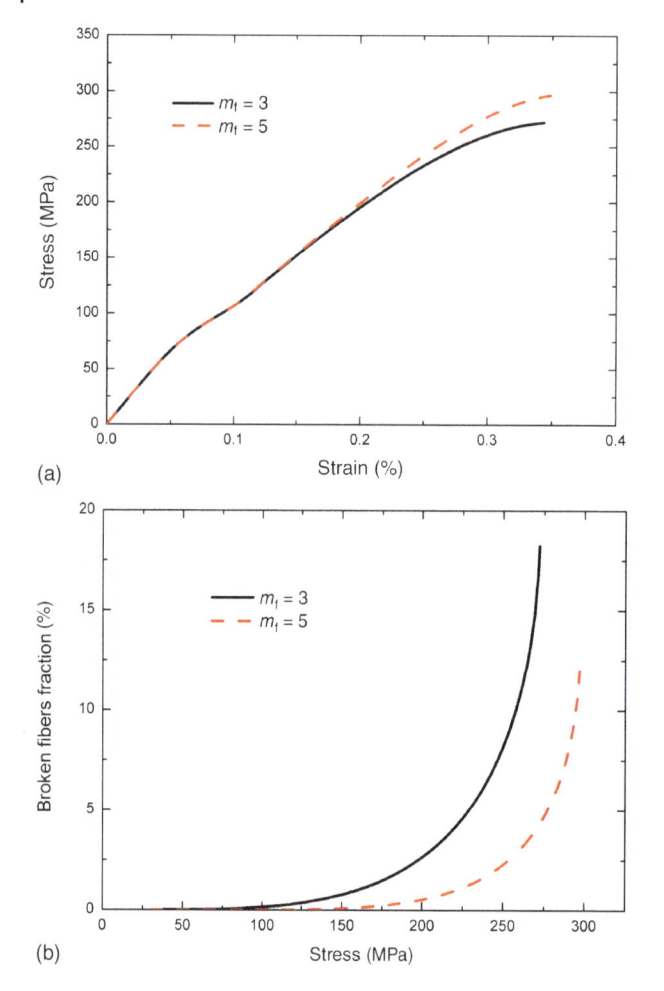

Figure 2.20 The effect of the fiber Weibull modulus on (a) the tensile stress–strain curves; and (b) the broken fibers fraction versus applied stress curves of C/SiC composite.

$\sigma_{UTS} = 264$ MPa with the failure strain of $\varepsilon_f = 0.61\%$, the interface debonding length increases to $2l_d/l_c = 1$, the interface oxidation ratio decreases to $\zeta/l_d = 49\%$, and the broken fibers fraction increases to $q = 12\%$; when the pre-exposure time is $t = 30$ hours, the composite tensile strength is $\sigma_{UTS} = 239$ MPa with the failure strain of $\varepsilon_f = 0.59\%$, the interface debonding length increases to $2l_d/l_c = 1$, the interface oxidation ratio decreases to $\zeta/l_d = 74.4\%$, and the broken fibers fraction increases to $q = 12.8\%$.

The experimental and predicted tensile stress–strain curves, interface debonding length and oxidation length, and fiber broken fraction of 2D C/SiC composite without and with pre-exposure at $T = 800\,°C$ and $t = 10, 20,$ and 30 hours are shown in Figure 2.22. With increasing of pre-exposure time, the composite tensile strength and failure strain both decrease; the interface debonded length and interface oxidation ratio increase; and the broken fibers fraction increases at low stress level. Without pre-exposure, the composite tensile strength is

2.3 Effect of Pre-exposure on Tensile Damage and Fracture of Ceramic-Matrix Composites

Table 2.2 The material properties of C/SiC composites.

Items	1D C/SiC	2D C/SiC	2.5D C/SiC	3D C/SiC
Λ	1	0.5	0.75	0.93
r_f (μm)	3.5	3.5	3.5	3.5
V_f (%)	30	35	40	40
E_f (GPa)	230	230	230	230
α_f (10^{-6}/K)	0	0	0.5	0
α_m (10^{-6}/K)	4.6	4.6	4.6	4.6
m	3	5	6	5
σ_R (MPa)	100	40	80	80
l_{sat} (μm)	120	300	80	80
τ_i (MPa)	10	11	5	9
τ_f (MPa)	1	1	1	1
ζ_d (J/m²)	0.1	0.3	0.1	0.1
σ_{UTS} (MPa)	333	149	226	206
ε_f (%)	0.59	0.34	0.56	0.37
m_f	5	5	5	5

$\sigma_{UTS} = 148$ MPa with the failure strain of $\varepsilon_f = 0.35\%$; when the pre-exposure time is $t = 10$ hours, the composite tensile strength is $\sigma_{UTS} = 148$ MPa with the failure strain of $\varepsilon_f = 0.37\%$, the interface debonding length increases to $2l_d/l_c = 38.4\%$, the interface oxidation ratio decreases to $\zeta/l_d = 8.6\%$, and the broken fibers fraction increases to $q = 11.7\%$; when the pre-exposure time is $t = 20$ hours, the composite tensile strength is $\sigma_{UTS} = 130$ MPa with the failure strain of $\varepsilon_f = 0.33\%$, the interface debonding length increases to $2l_d/l_c = 35.4\%$, the interface oxidation ratio decreases to $\zeta/l_d = 18.6\%$, and the broken fibers fraction increases to $q = 12.2\%$; when the pre-exposure time is $t = 30$ hours, the composite tensile strength is $\sigma_{UTS} = 117$ MPa with the failure strain of $\varepsilon_f = 0.3\%$, the interface debonding length increases to $2l_d/l_c = 34\%$, the interface oxidation ratio decreases to $\zeta/l_d = 29\%$, and the broken fibers fraction increases to $q = 10.9\%$.

The experimental and predicted tensile stress–strain curves, interface debonded length and oxidation length, and fiber broken fraction of 2.5D C/SiC composite without and with pre-exposure at $T = 900\,°C$ and $t = 10$ hours are shown in Figure 2.23. Without pre-exposure, the composite tensile strength is $\sigma_{UTS} = 225$ MPa with the failure strain of $\varepsilon_f = 0.54\%$; when the pre-exposure time is $t = 10$ hours, the composite tensile strength is $\sigma_{UTS} = 191$ MPa with the failure strain of $\varepsilon_f = 0.48\%$, the interface debonding length increases to $2l_d/l_c = 1$, the interface oxidation ratio decreases to $\zeta/l_d = 25\%$, and the broken fibers fraction increases to $q = 23.5\%$.

The experimental and predicted tensile stress–strain curves, interface debonded length and oxidation length, and fiber broken fraction of 3D C/SiC

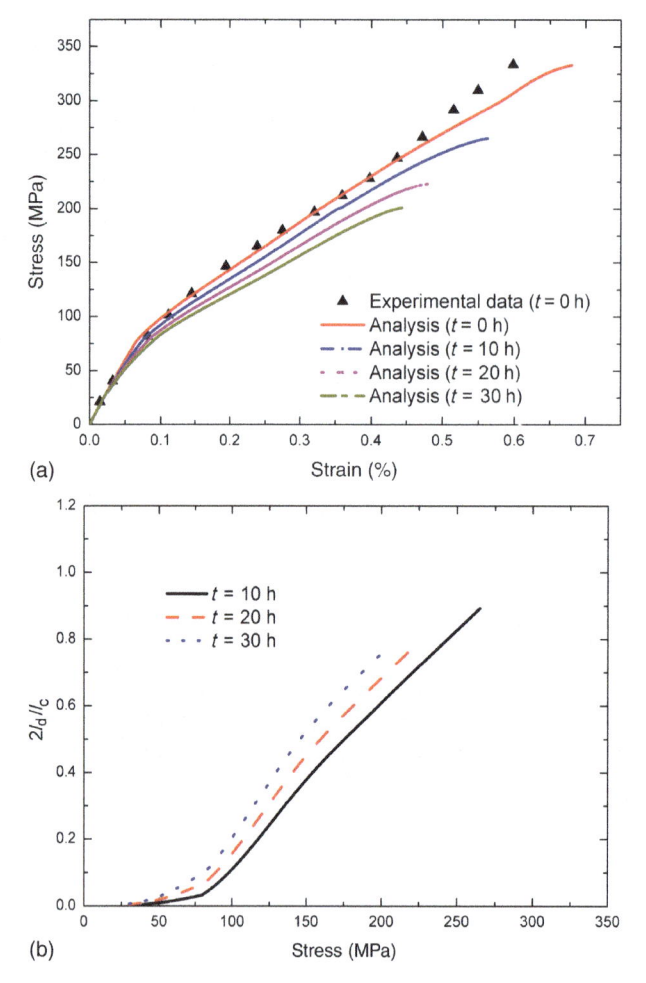

Figure 2.21 (a) The experimental and predicted tensile stress–strain curves; (b) the interface debonding length of $2l_d/l_c$ versus the applied stress curves; (c) the interface oxidation ratio of ζ/l_d versus the applied stress curves; and (d) the broken fibers fraction of q versus applied stress curves of unidirectional C/SiC composite.

composite without and with pre-exposure at $T = 800\,°C$ and $t = 10, 20,$ and 30 hours are shown in Figure 2.24. With increasing of pre-exposure time, the composite tensile strength and failure strain both decrease; the interface debonded length and interface oxidation ratio increase; and the broken fibers fraction increases at low stress level. Without pre-exposure, the composite tensile strength is $\sigma_{UTS} = 203$ MPa with the failure strain of $\varepsilon_f = 0.38\%$; when the pre-exposure time is $t = 10$ hours, the composite tensile strength is $\sigma_{UTS} = 192$ MPa with the failure strain of $\varepsilon_f = 0.38\%$, the interface debonding length increases to $2l_d/l_c = 83\%$, the interface oxidation ratio decreases to $\zeta/l_d = 15\%$, and the broken fibers fraction increases to $q = 13.1\%$; when the pre-exposure time is $t = 20$ hours, the composite tensile

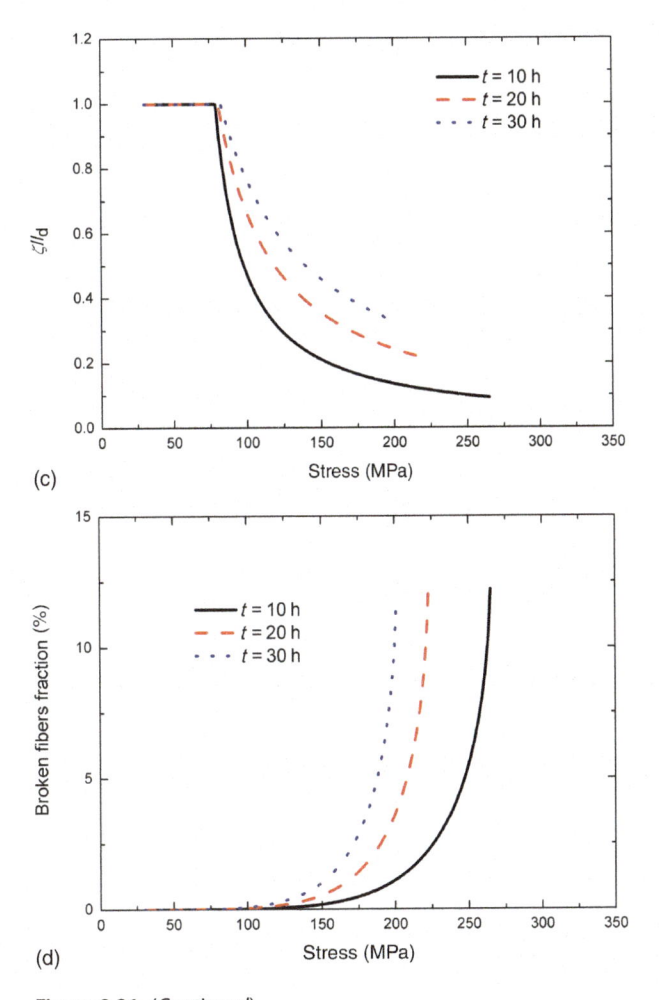

Figure 2.21 (Continued)

strength is $\sigma_{UTS} = 161$ MPa with the failure strain of $\varepsilon_f = 0.33\%$, the interface debonding length increases to $2l_d/l_c = 79\%$, the interface oxidation ratio decreases to $\zeta/l_d = 31\%$, and the broken fibers fraction increases to $q = 11.7\%$; when the pre-exposure time is $t = 30$ hours, the composite tensile strength is $\sigma_{UTS} = 145$ MPa with the failure strain of $\varepsilon_f = 0.32\%$, the interface debonding length increases to $2l_d/l_c = 83\%$, the interface oxidation ratio decreases to $\zeta/l_d = 44\%$, and the broken fibers fraction increases to $q = 10.8\%$.

2.4 Effect of Interface Properties on Lifetime of Ceramic-Matrix Composites

In this section, the fatigue life of fiber-reinforced CMCs with different fiber preforms, i.e. unidirectional, cross-ply, 2D, 2.5D, and 3D CMCs at room and elevated

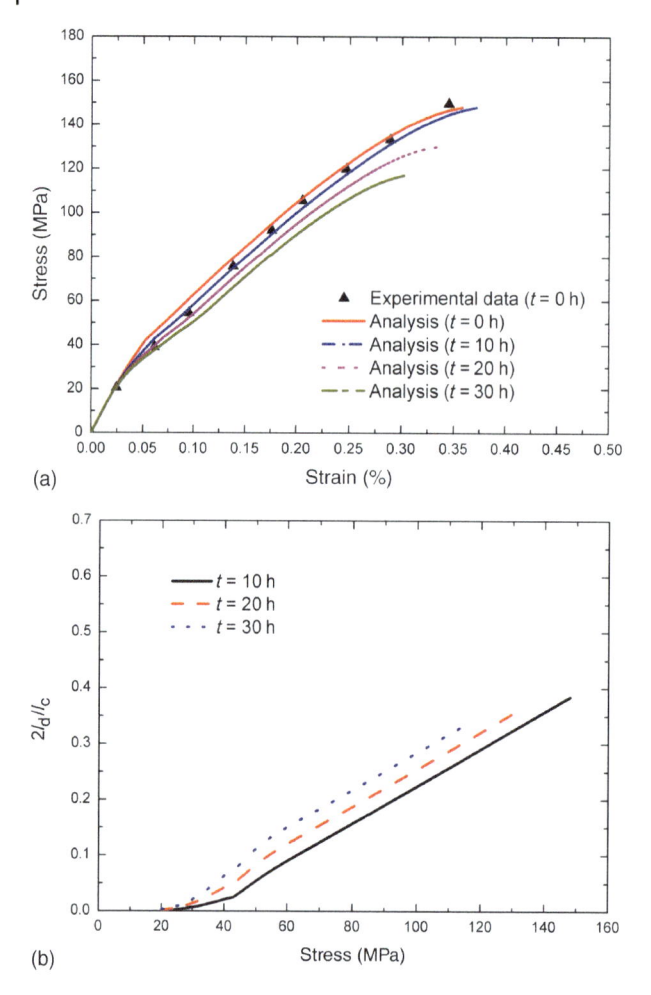

Figure 2.22 (a) The experimental and predicted tensile stress–strain curves; (b) the interface debonding length of $2l_d/l_c$ versus the applied stress curves; (c) the interface oxidation ratio of ζ/l_d versus the applied stress curves; and (d) the broken fibers fraction of q versus applied stress curves of 2D C/SiC composite.

temperatures in air and oxidative environment, are predicted using the micromechanics approach. An ECFL is introduced to describe the fiber architecture of preforms. Under cyclic fatigue loading, the fiber broken fraction is determined by combining the interface wear model and fiber statistical failure model at room temperature, and interface/fiber oxidation model, interface wear model and fiber statistical failure model at elevated temperature, based on the assumption that the fiber strength is subjected to two-parameter Weibull distribution and the load carried by broken and intact fibers satisfy the GLS criterion. When the broken fibers fraction approaches the critical value, the composites fatigue fracture.

(c)

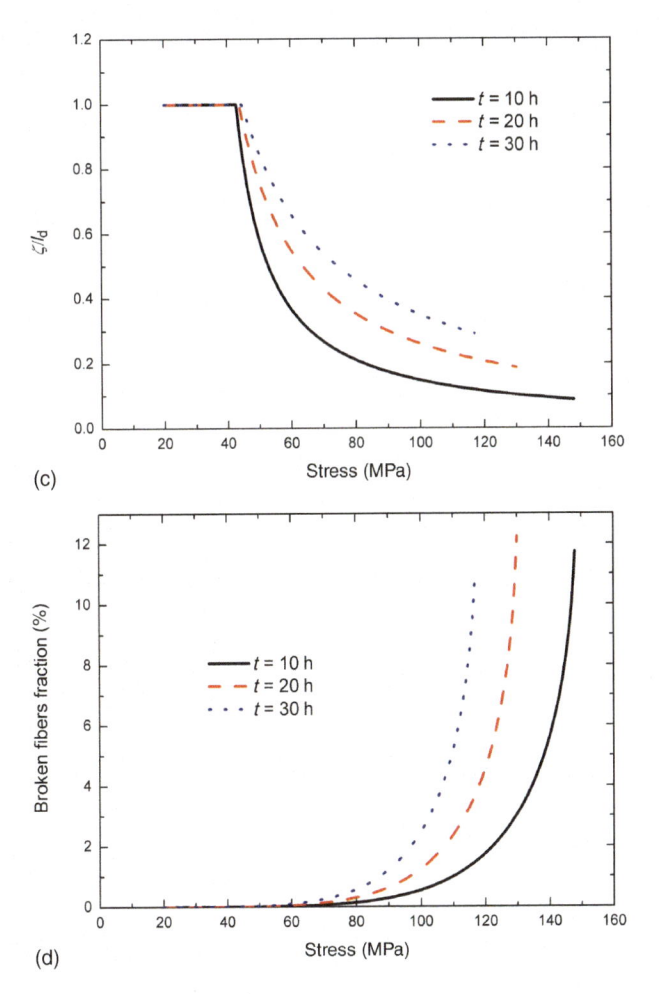

(d)

Figure 2.22 (Continued)

2.4.1 Theoretical Analysis

When fibers begin to break, the load dropped by broken fibers would be transferred to intact fibers in the cross-section. Two dominant failure criterions are present in the literatures for modeling fiber failure, i.e. GLS criterion (Curtin 2000) and Local Load Sharing (LLS) criterion (Xia and Curtin 2000). The GLS criterion assumes that the load from any one fiber is transferred equally to all other intact fibers at the same cross-section plane. The GLS assumption neglects any local stress concentrations in the neighborhood of existing breaks and is expected to be accurate when the interface shear stress is sufficiently low. The two-parameter Weibull model is adopted to describe the fibers strength distribution. The fiber fracture probability q is (Curtin et al. 1998)

$$q = 1 - \exp\left(-\int_{L_0} \frac{1}{l_0}\left[\frac{\sigma_f(x)}{\sigma_0}\right]^{m_f} dx\right) \tag{2.46}$$

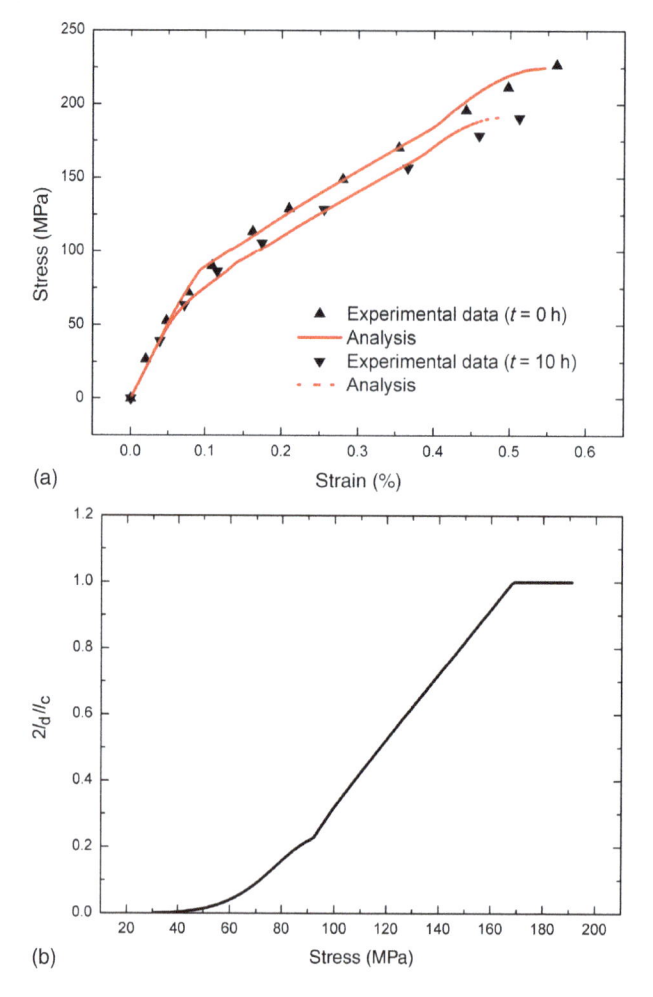

Figure 2.23 (a) The experimental and predicted tensile stress–strain curves; (b) the interface debonded length of $2l_d/l_c$ versus the applied stress curves; (c) the interface oxidation ratio of ζ/l_d versus the applied stress curves; and (d) the broken fibers fraction of q versus applied stress curves of 2.5D C/SiC composite.

where σ_0 denotes the fiber strength at tested gauge length of l_0; m_f denotes the fiber Weibull modulus; and L_0 denotes the integral length.

2.4.1.1 Life Prediction Model at Room Temperature

The GLS assumption is used to determine the load carried by intact and fracture fibers.

$$\frac{\sigma}{V_f} = \Phi[1 - q(\Phi)] + \langle \Phi_b \rangle q(\Phi) \tag{2.47}$$

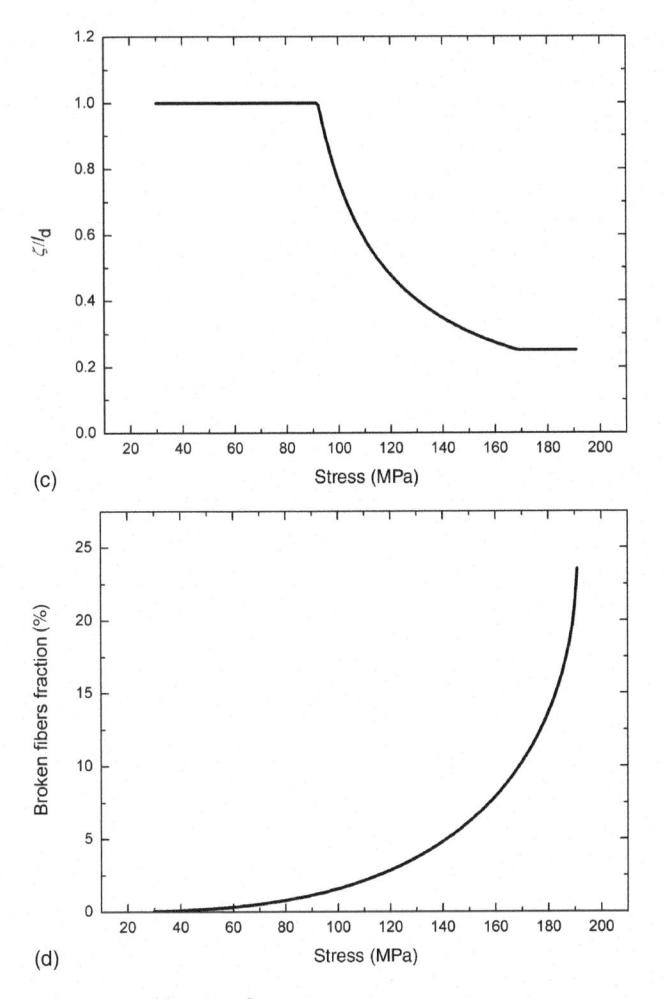

Figure 2.23 (Continued)

where $\langle \Phi_b \rangle$ denotes the average stress carried by broken fibers.

$$\langle \Phi_b \rangle = \frac{\Phi}{q(\Phi)} \left(\frac{\sigma_c}{\Phi}\right)^{m_f+1} \left(\frac{\sigma_0(N)}{\sigma_0}\right)^{m_f} \frac{\tau_i(N)}{\tau_i} \left\{ 1 - \exp\left[-\left(\frac{\Phi}{\sigma_c}\right)^{m_f+1} \left(\frac{\sigma_0}{\sigma_0(N)}\right)^{m_f} \frac{\tau_i}{\tau_i(N)}\right] \right\} - \frac{\Phi}{q(\Phi)} \exp\left\{-\left(\frac{\Phi}{\sigma_c}\right)^{m_f+1} \left(\frac{\sigma_0}{\sigma_0(N)}\right)^{m_f} \frac{\tau_i}{\tau_i(N)}\right\} \quad (2.48)$$

and

$$q(\Phi) = 1 - \exp\left\{-\left(\frac{\Phi}{\sigma_c}\right)^{m_f+1} \left(\frac{\sigma_0}{\sigma_0(N)}\right)^{m_f} \frac{\tau_i}{\tau_i(N)}\right\} \quad (2.49)$$

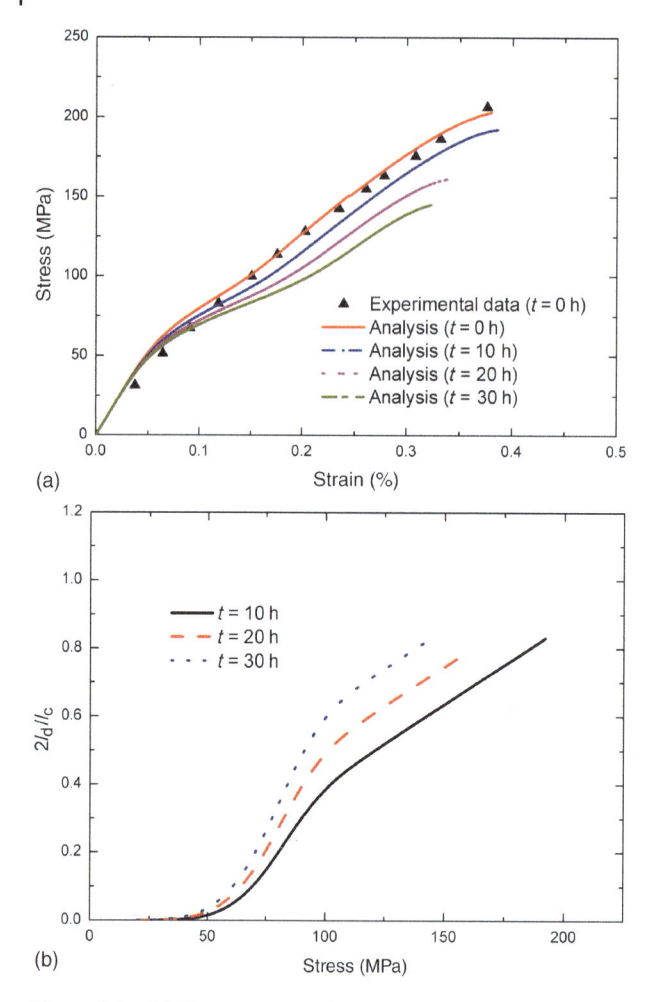

2 Interface Characterization of Ceramic-Matrix Composites

Figure 2.24 (a) The experimental and predicted tensile stress–strain curves; (b) the interface debonding length of $2l_d/l_c$ versus the applied stress curves; (c) the interface oxidation ratio of ζ/l_d versus the applied stress curves; and (d) the broken fibers fraction of q versus applied stress curves of 3D C/SiC composite.

Substituting Eqs. (2.48) and (2.49) into Eq. (2.47), it leads to the form of

$$\frac{\sigma}{V_f} = \Phi \left(\frac{\sigma_c}{\Phi}\right)^{m_f+1} \left(\frac{\sigma_0(N)}{\sigma_0}\right)^{m_f} \frac{\tau_i(N)}{\tau_i} \left\{ 1 - \exp\left[-\left(\frac{\Phi}{\sigma_c}\right)^{m_f+1}\right.\right.$$
$$\left.\left.\left(\frac{\sigma_0}{\sigma_0(N)}\right)^{m_f} \frac{\tau_i}{\tau_i(N)}\right]\right\} \tag{2.50}$$

where

$$\tau_i(N) = \tau_{io} + [1 - \exp(-\omega N^\lambda)](\tau_{imin} - \tau_{io}) \tag{2.51}$$

$$\sigma_0(N) = \sigma_0[1 - p_1(\log N)^{p_2}] \tag{2.52}$$

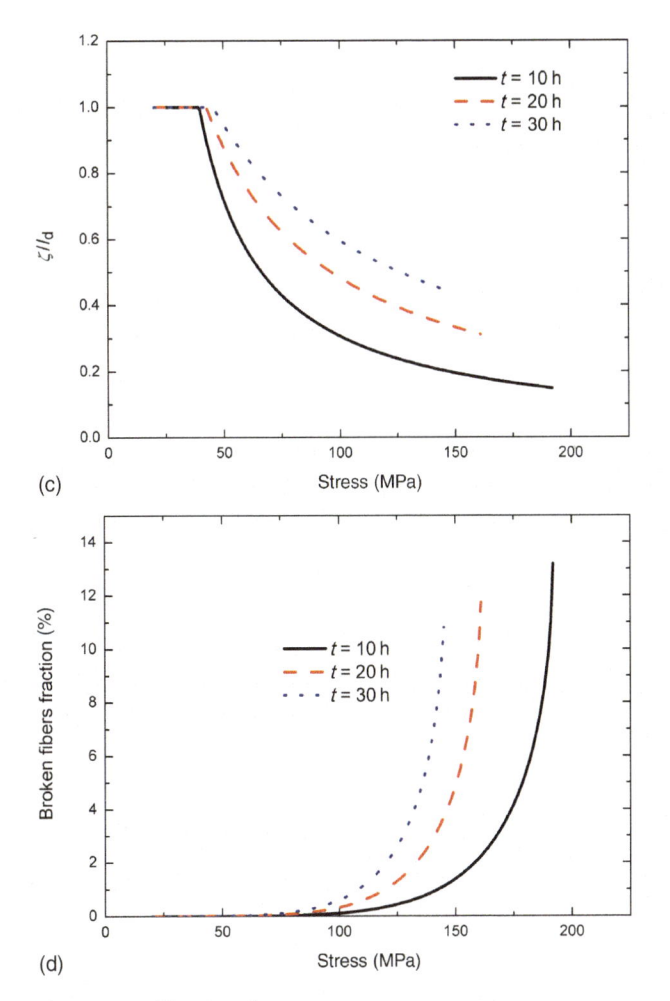

Figure 2.24 (Continued)

Using Eqs. (2.50)–(2.52), the stress Φ carried by intact fibers at the matrix cracking plane can be determined for different fatigue peak stresses. Substituting Eqs. (2.51) and (2.52) and the intact fibers stress Φ into Eq. (2.49), the fiber failure probability corresponding to different cycle number can be determined. When the broken fibers fraction approaches the critical value, the composites fatigue fracture.

2.4.1.2 Life Prediction Model at Elevated Temperatures in the Oxidative Environment

When fiber-reinforced CMCs subjected to oxidation, the notch would form at the fiber surface leading to the degradation of fiber strength and the increase of fiber stress concentration and fracture probability. The fracture probabilities of oxidized fibers in the oxidation region, unoxidized fibers in the oxidation region, fibers in the interface debonded region, and interface bonded region of $q_a(\Phi)$,

$q_b(\Phi)$, $q_c(\Phi)$, and $q_d(\Phi)$ are

$$q_a(\Phi) = 1 - \exp\left\{-2\frac{\zeta(t)}{l_0}\left[\frac{\Phi}{\sigma_0(t)}\right]^{m_f}\right\} \tag{2.53}$$

$$q_b(\Phi) = 1 - \exp\left\{-2\frac{\zeta(t)}{l_0}\left(\frac{\Phi}{\sigma_0}\right)^{m_f}\right\} \tag{2.54}$$

$$q_c(\Phi) = 1 - \exp\left\{-\frac{r_f \Phi^{m_f+1}}{l_0(\sigma_0(N))^{m_f}\tau_i(N)(m_f+1)}\left[1-\left(1-\frac{l_d(N)}{l_f(N)}\right)^{m_f+1}\right]\right\} \tag{2.55}$$

$$q_d(\Phi) = 1 - \exp\left\{-\frac{2r_f\Phi^{m_f}}{\rho l_0(\sigma_0(N))^{m_f}(m_f+1)\left(1-\frac{\sigma_{fo}}{\Phi}-\frac{l_d(N)}{l_s(N)}\right)}\right.$$
$$\times\left[\left(1-\frac{l_d(N)}{l_f(N)}-\left(1-\frac{\sigma_{fo}}{\Phi}-\frac{l_d(N)}{l_f(N)}\right)\frac{\rho l_d(N)}{r_f}\right)^{m_f+1}\right.$$
$$\left.\left.-\left(1-\frac{l_d(N)}{l_f(N)}-\left(1-\frac{\sigma_{fo}}{\Phi}-\frac{l_d(N)}{l_f(N)}\right)\frac{\rho l_c}{2r_f}\right)^{m_f+1}\right]\right\} \tag{2.56}$$

where l_f denotes the slip length over which the fiber stress would decay to zero if not interrupted by the far-field equilibrium stresses.

$$l_f(N) = \frac{r_f \Phi}{2\tau_i(N)} \tag{2.57}$$

$$\sigma_0(t) = \begin{cases} \sigma_0, & t \leq \frac{1}{k}\left(\frac{K_{IC}}{Y\sigma_0}\right)^4 \\ \frac{K_{IC}}{Y\sqrt[4]{kt}}, & t > \frac{1}{k}\left(\frac{K_{IC}}{Y\sigma_0}\right)^4 \end{cases} \tag{2.58}$$

The GLS assumption is used to determine the load carried by intact and fracture fibers (Curtin et al. 1998).

$$\frac{\sigma}{V_f} = \left[1-q_f(\Phi)\left(1+\frac{2l_f}{l_c}\right)\right]\Phi + q_r(\Phi)\frac{2l_f}{l_c}\langle\Phi_b\rangle \tag{2.59}$$

where

$$q_f(\Phi) = \varphi[\eta q_a(\Phi) + (1-\eta)q_b(\Phi)] + q_c(\Phi) + q_d(\Phi) \tag{2.60}$$

$$q_r(\Phi) = q_c(\Phi) + q_d(\Phi) \tag{2.61}$$

where η denotes the oxidation fibers fraction in the oxidized region; and φ denotes the fraction of oxidation in the multiple matrix cracks.

$$\varphi = \frac{l_{sat}}{l_f - 2\zeta(t)} \tag{2.62}$$

The average stress carried by broken fibers is given by Eq. (2.62).

$$\langle \Phi_b \rangle = \int_0^{l_f} \Phi_b(x) f(x) \, dx$$

$$= \frac{\Phi}{q_r(\Phi)} \left(\frac{\sigma_c}{\Phi}\right)^{m_f+1} \left(\frac{\sigma_0(N)}{\sigma_0}\right)^{m_f} \frac{\tau_i(N)}{\tau_i} \left\{ 1 - \exp\left[-\left(\frac{\Phi}{\sigma_c}\right)^{m_f+1} \left(\frac{\sigma_0}{\sigma_0(N)}\right)^{m_f} \frac{\tau_i}{\tau_i(N)} \right] \right\}$$

$$- \frac{\Phi}{P_r(\Phi)} \exp\left\{ -\left(\frac{\Phi}{\sigma_c}\right)^{m_f+1} \left(\frac{\sigma_0}{\sigma_0(N)}\right)^{m_f} \frac{\tau_i}{\tau_i(N)} \right\} \quad (2.63)$$

Substituting Eqs. (2.60), (2.61), and (2.63) into Eq. (2.59), the stress Φ carried by intact fibers at the matrix crack plane can be determined for different cycle number and fatigue stress. Substituting Eqs. (2.51), (2.52), (2.58), and (2.16) and the intact fiber stress Φ into Eqs. (2.60) and (2.61), the fiber failure probabilities corresponding to different number of applied cycles can be determined. When the broken fibers fraction approaches the critical value, the composites fatigue fracture.

2.4.2 Experimental Comparisons

Under cyclic fatigue loading, the loading directions were along with fiber for the unidirectional CMCs, 0° fiber ply for the cross-ply and plain-weave 2D CMCs, warp yarn for the 2.5D CMCs, and axial fibers at a small angle θ for 3D CMCs. An ECFL is defined as

$$\Lambda = \frac{V_{f\_axial}}{V_f} \quad (2.64)$$

where V_f and V_{f_axial} denote the total fiber volume fraction in the composites and the effective fiber volume fraction in the cyclic loading direction, respectively. Under cyclic fatigue loading at room and elevated temperatures, the broken fibers fraction in the 0° plies or longitudinal yarns of cross-ply and 2.5D CMCs would increase with the increase of loading cycles and oxidation time. When the broken fibers fraction in the 0° plies or longitudinal yarns approaches the critical value, the composite would be fatigue failure.

2.4.2.1 Life Prediction at Room Temperature

The tensile strength of unidirectional C/SiC composite is $\sigma_{UTS} = 270$ MPa, and the fatigue peak stresses are 0.51, 0.66, 0.74, 0.88, and 0.96 of tensile strength; the tensile strength of unidirectional SiC/CAS composite is $\sigma_{UTS} = 450$ MPa, and the fatigue peak stresses are 0.8, 0.72, and 0.7 of tensile strength (Evans et al. 1995); the tensile strength of unidirectional SiC/1723 composite is $\sigma_{UTS} = 680$ MPa, and the fatigue peak stresses are 0.51, 0.58, 0.66, 0.70, 0.72, 0.76, and 0.77 of tensile strength (Zawada et al. 1991); the tensile strength of cross-ply C/SiC composite is $\sigma_{UTS} = 124$ MPa, and the fatigue peak stresses are 0.70, 0.80, 0.85, and 0.90 of tensile strength; the tensile strength of cross-ply SiC/CAS composite is $\sigma_{UTS} = 275$ MPa, and the fatigue peak stresses are 0.5, 0.58, and 0.65 of tensile strength (Opalski and Mall 1994); the tensile strength of cross-ply SiC/1723 composite is $\sigma_{UTS} = 284$ MPa, and the fatigue peak stresses are 0.6, 0.7, and 0.74 of tensile strength (Zawada et al. 1991); the tensile strength of 2D SiC/SiC composite is $\sigma_{UTS} = 170$ MPa, and the fatigue peak stresses are 0.80, 0.81, 0.82, 0.84, 0.87, and 0.94 of tensile strength (Rouby and Reynaud 1993); the tensile strength of 2D C/SiC composite is $\sigma_{UTS} = 420$ MPa, and the fatigue peak stresses are 0.80, 0.83, 0.86, 0.89, 0.91, and 0.96 of tensile strength (Shuler et al. 1993); the tensile strength of 2.5D C/SiC composite is $\sigma_{UTS} = 225$ MPa, and the fatigue peak stresses are 0.6, 0.7, 0.75, and 0.8 of tensile strength (Yang 2011); and the tensile strength of 3D C/SiC is $\sigma_{UTS} = 276$ MPa, and the fatigue peak stress are 0.80, 0.83, 0.87, 0.89, 0.90, and 0.94 of tensile strength (Du et al. 2002).

For unidirectional C/SiC composite, the interface shear stress versus applied cycles curve is simulated by the Evans–Zok–McMeeking model (Evans et al. 1995), as shown in Figure 2.25a, in which the model parameters are given by $\tau_{io} = 8$ MPa, $\tau_{imin} = 0.3$ MPa, $\omega = 0.04$, and $\lambda = 1.5$. The material properties are given by $V_f = 40\%$, $\Lambda = 1.0$, $r_f = 3.5$ μm, $m_f = 5$, $p_1 = 0.02$, and $p_2 = 1.0$. The broken fibers fraction versus cycle number curves under $\sigma_{max} = 267$ and 260 MPa are shown in Figure 2.25b. Under the fatigue peak stress of $\sigma_{max} = 267$ MPa, the composite fatigue failed after 31 cycles; and under the fatigue peak stress of $\sigma_{max} = 260$ MPa, the composite fatigue failed after 400 cycles. The experimental and theoretical fatigue life S–N curves are shown in Figure 2.25c, in which the fatigue limit approaches 88% of tensile strength.

For unidirectional SiC/CAS composite, the interface shear stress versus applied cycles curve is simulated using the Evans–Zok–McMeeking model (Evans et al. 1995), as shown in Figure 2.26a, in which the model parameters are given by $\tau_{io} = 22$ MPa, $\tau_{imin} = 5$ MPa, $\omega = 0.001$, and $\lambda = 2$. The material properties are given by $V_f = 38\%$, $\Lambda = 1.0$, $r_f = 7.5$ μm, $m_f = 2$, $p_1 = 0.02$, and $p_2 = 1.0$. The broken fibers fraction versus applied cycle number curves under the fatigue peak stresses of $\sigma_{max} = 420$, 380, 340, and 320 MPa are shown in Figure 2.26b. Under the fatigue peak stresses of $\sigma_{max} = 420$ and 380 MPa, the composite fatigue failed after 43 and 62 cycles, respectively. Under the fatigue peak stresses of $\sigma_{max} = 340$ and 320 MPa, the composite failed after 2005 and 14 788 cycles, respectively. The experimental and theoretical fatigue life S–N curves are shown in Figure 2.26c, in which the fatigue limit approaches 62% of tensile strength.

For unidirectional SiC/1723 composite, the interface shear stress versus applied cycles curve is simulated using the Evans–Zok–McMeeking model (Evans et al. 1995), as shown in Figure 2.27a, in which the model parameters are given by $\tau_{io} = 32$ MPa, $\tau_{imin} = 27$ MPa, $\omega = 0.001$, and $\lambda = 1.5$. The material properties are given by $V_f = 45\%$, $\Lambda = 1.0$, $r_f = 6.25$ μm, $m_f = 2$, $p_1 = 0.02$, and $p_2 = 1.25$. The broken fibers fraction versus cycle number curves under the fatigue peak stresses of $\sigma_{max} = 650$, 600, 550, 500, and 480 MPa are shown in Figure 2.27b. When the broken fibers fraction approaches the critical value of 37%, the composite would fatigue fail. The corresponding failure cycles are 2872, 15 026, 70 823, 305 366, and 535 575 cycles, respectively. The experimental and theoretical fatigue life S–N curves are shown in Figure 2.27c, in which the fatigue limit approaches 70% of tensile strength.

For cross-ply C/SiC composite, the interface shear stress versus applied cycles curve is simulated by the Evans–Zok–McMeeking model (Evans et al. 1995), as shown in Figure 2.28a, in which the model parameters are given by $\tau_{io} = 6.2$ MPa,

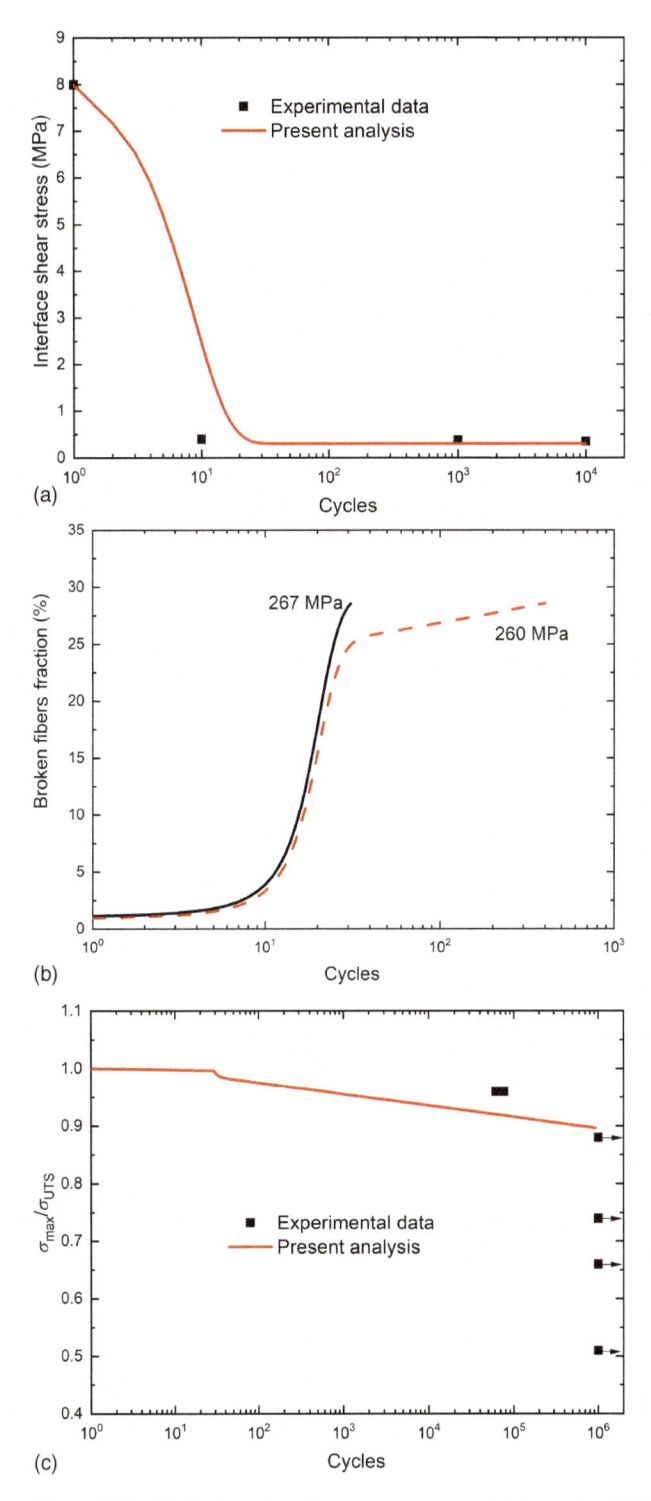

Figure 2.25 (a) The interface shear stress versus applied cycles; (b) the broken fibers fraction versus applied cycles; and (c) the fatigue life S–N curves of experimental data and theoretical analysis for unidirectional C/SiC composite at room temperature.

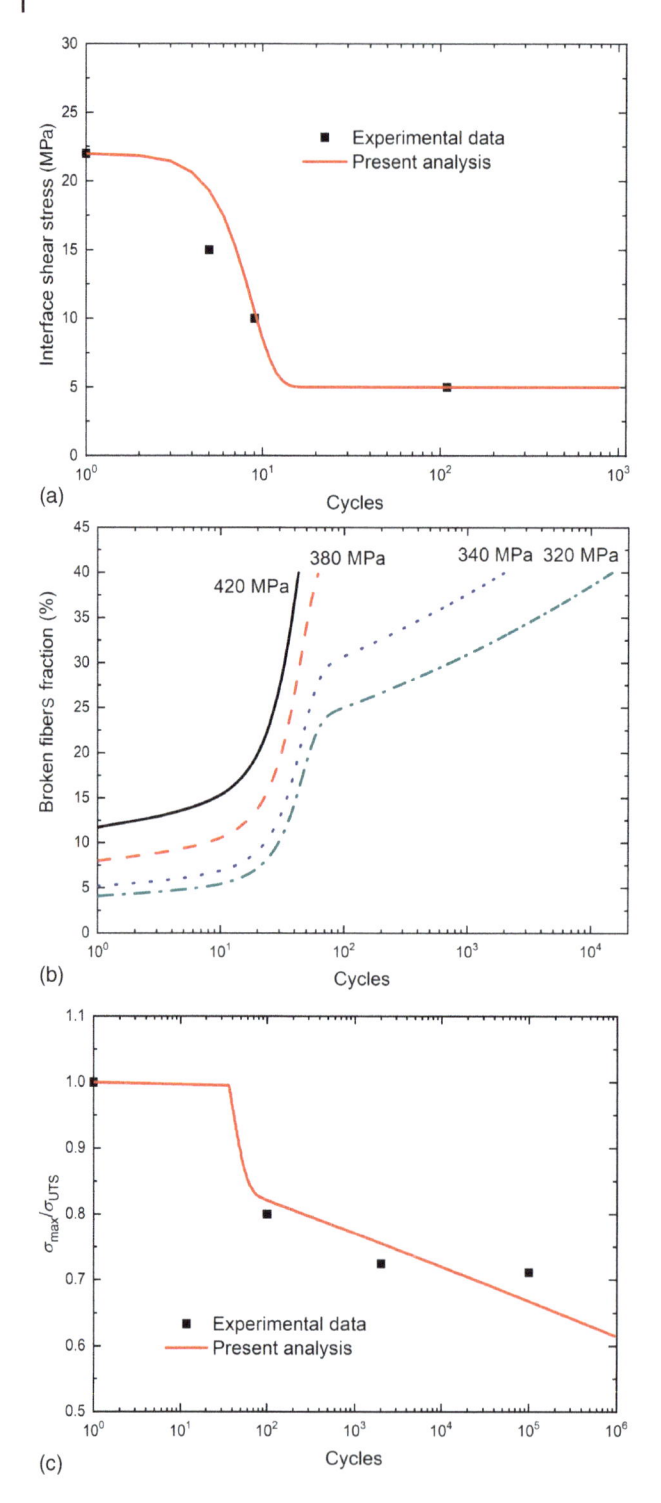

Figure 2.26 (a) The interface shear stress versus applied cycles; (b) the broken fibers fraction versus applied cycles; and (c) the fatigue life S–N curves of experimental data and theoretical analysis for unidirectional SiC/CAS composite at room temperature.

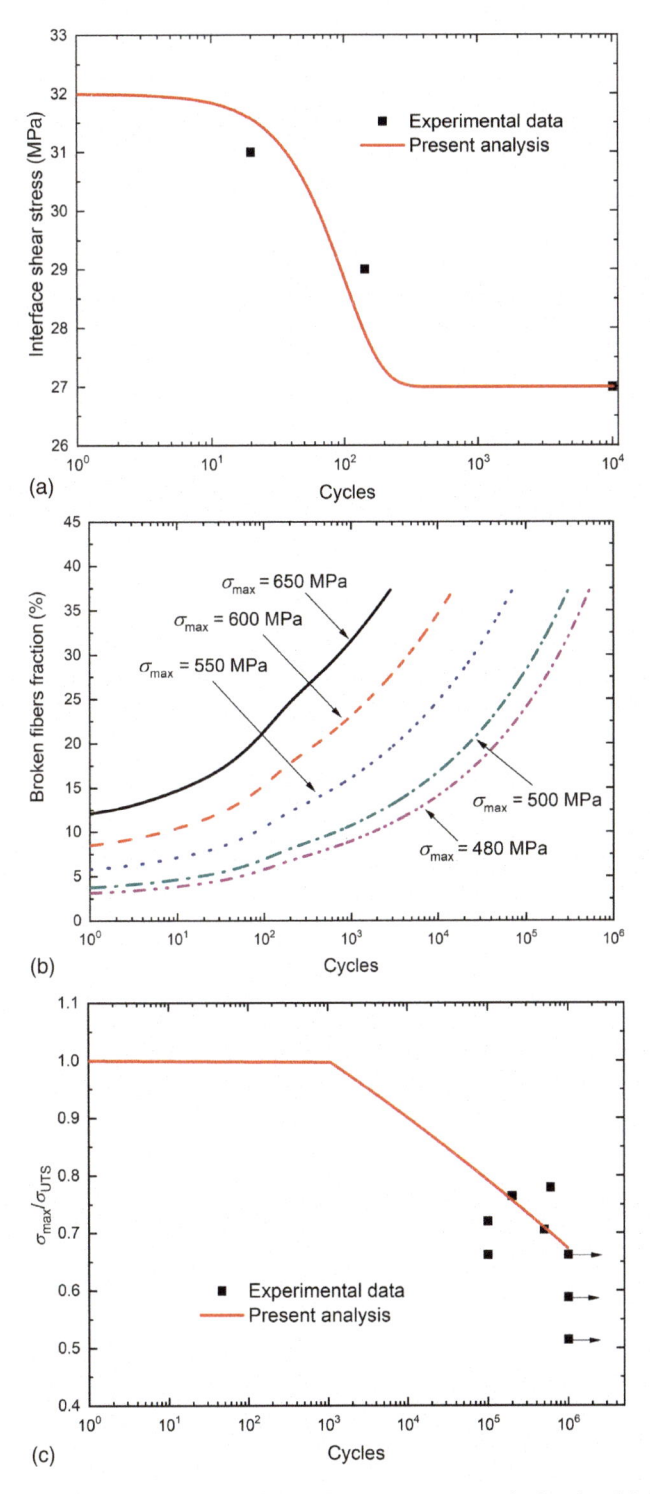

Figure 2.27 (a) The interface shear stress versus applied cycles; (b) the broken fibers fraction versus applied cycles; and (c) the fatigue life S–N curves of experimental data and theoretical analysis for unidirectional SiC/1723 composite at room temperature.

$\tau_{\text{imin}} = 1.5\,\text{MPa}$, $\omega = 0.06$, and $\lambda = 1.8$. The material properties are given by $V_f = 40\%$, $\Lambda = 0.5$, $r_f = 3.5\,\mu\text{m}$, $m_f = 5$, $p_1 = 0.01$, and $p_2 = 0.8$. The broken fibers fraction versus cycle number curves under the fatigue peak stresses of $\sigma_{\max} = 110$ and $108\,\text{MPa}$ are shown in Figure 2.28b. Under the fatigue peak stress of $\sigma_{\max} = 110\,\text{MPa}$, the composite fatigue failed after 10 cycles; and under the fatigue peak stress of $\sigma_{\max} = 108\,\text{MPa}$, the composite fatigue failed after 53 cycles. The experimental and theoretical fatigue life S–N curves are shown in Figure 2.28c, in which the fatigue limit approaches 88% of tensile strength.

For cross-ply SiC/CAS composite, the interface shear stress versus applied cycles curve is simulated using the Evans–Zok–McMeeking model (Evans et al. 1995), as shown in Figure 2.29a, in which the model parameters are given by $\tau_{io} = 22\,\text{MPa}$, $\tau_{\text{imin}} = 5\,\text{MPa}$, $\omega = 0.001$, and $\lambda = 2$. The material properties are given by $V_f = 35\%$, $\Lambda = 0.5$, $r_f = 7.5\,\mu\text{m}$, $m_f = 2$, $p_1 = 0.05$, and $p_2 = 1.0$. The broken fibers fraction versus cycle number curves under the fatigue peak stresses of $\sigma_{\max} = 240$, 200, and $180\,\text{MPa}$ are shown in Figure 2.29b. Under the fatigue peak stress of $\sigma_{\max} = 240\,\text{MPa}$, the composite fatigue failed after 291 cycles; and under the fatigue peak stresses of $\sigma_{\max} = 200$ and $180\,\text{MPa}$, the composite fatigue failed after 3442 and 19 477 cycles, respectively. The experimental and theoretical fatigue life S–N curves are shown in Figure 2.29c, in which the fatigue limit approaches 48% of tensile strength.

For cross-ply SiC/1723 composite, the interface shear stress versus applied cycles curve is simulated using the Evans–Zok–McMeeking model (Evans et al. 1995), as shown in Figure 2.30a, in which the model parameters are given by $\tau_{io} = 32\,\text{MPa}$, $\tau_{\text{imin}} = 27\,\text{MPa}$, $\omega = 0.001$, and $\lambda = 1.5$. The material properties are given by $V_f = 45\%$, $\Lambda = 0.5$, $r_f = 6.25\,\mu\text{m}$, $m_f = 2$, $p_1 = 0.02$, and $p_2 = 1.25$. The broken fibers fraction versus cycle number curves under the fatigue peak stresses of $\sigma_{\max} = 280$, 260, and $240\,\text{MPa}$ are shown in Figure 2.30b. When the broken fibers fraction approaches the critical value of 35%, the composite would fatigue fail. The corresponding failure cycles are 1053, 5619, and 26 830 cycles, respectively. The experimental and theoretical fatigue life S–N curves are shown in Figure 2.30c, in which the fatigue limit approaches 67% of tensile strength.

For 2D SiC/SiC composite, the interface shear stress versus applied cycles curve is simulated using the Evans–Zok–McMeeking model (Evans et al. 1995), as shown in Figure 2.31a, in which the model parameters are given by $\tau_{io} = 56\,\text{MPa}$, $\tau_{\text{imin}} = 5\,\text{MPa}$, $\omega = 0.04$, and $\lambda = 1.0$. The material properties are given by $V_f = 40\%$, $\Lambda = 0.5$, $r_f = 7.5\,\mu\text{m}$, $m_f = 2$, $p_1 = 0.025$, and $p_2 = 1.0$. The broken fibers fraction versus cycle number curves under the fatigue peak stresses of $\sigma_{\max} = 160$, 150, and $140\,\text{MPa}$ are shown in Figure 2.31b. Under the fatigue peak stress of $\sigma_{\max} = 160\,\text{MPa}$, the composite fatigue failed after 122 cycles; and under the fatigue peak stresses of $\sigma_{\max} = 150$ and $140\,\text{MPa}$, the composite failed after 933 and 24 310 cycles, respectively. The experimental and theoretical fatigue life S–N curves are illustrated in Figure 2.31c, in which the fatigue limit approaches 78% of tensile strength.

For 2D C/SiC composite, the interface shear stress versus applied cycles curve is simulated using the Evans–Zok–McMeeking model (Evans et al. 1995), as shown in Figure 2.32a, in which the model parameters are given by $\tau_{io} = 25\,\text{MPa}$, $\tau_{\text{imin}} = 8\,\text{MPa}$, $\omega = 0.002$, and $\lambda = 1.0$. The material properties are given by

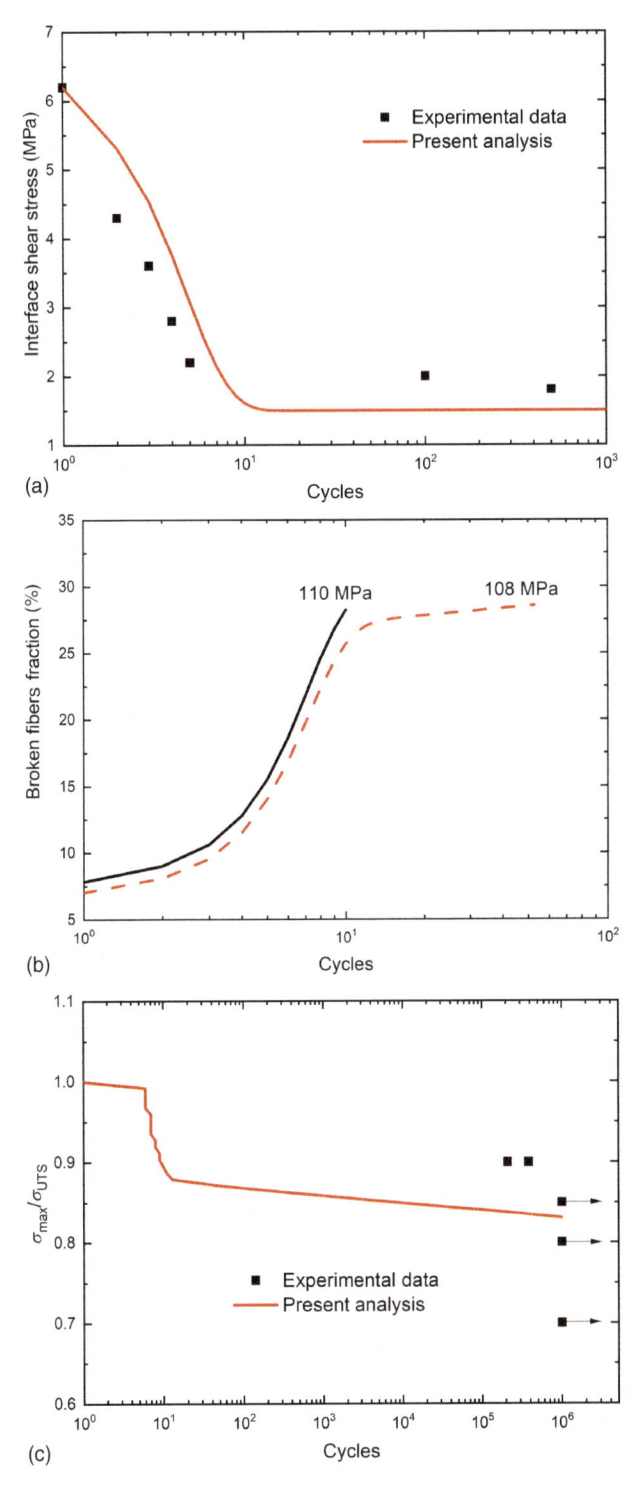

2.4 Effect of Interface Properties on Lifetime of Ceramic-Matrix Composites | 85

Figure 2.28 (a) The interface shear stress versus applied cycles; (b) the broken fibers fraction versus applied cycles; and (c) the fatigue life S–N curves of experimental data and theoretical analysis for cross-ply C/SiC composite at room temperature.

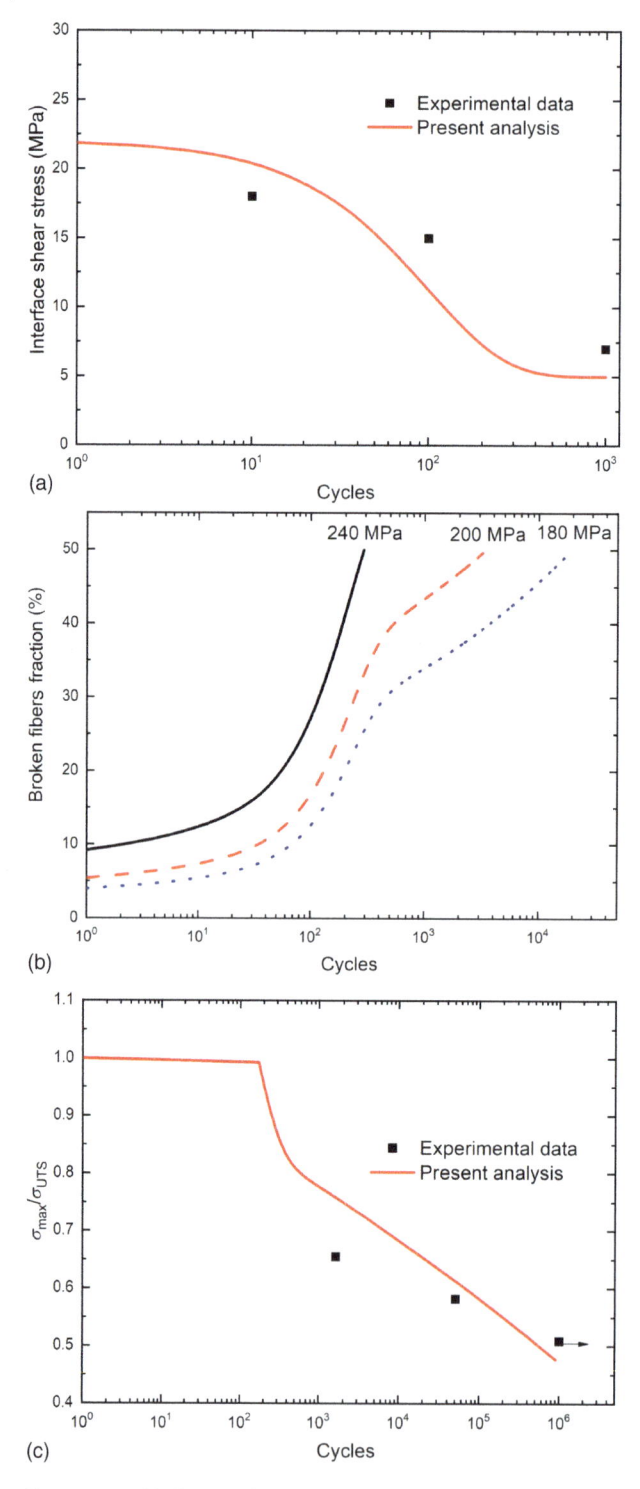

Figure 2.29 (a) The interface shear stress versus applied cycles; (b) the broken fibers fraction versus applied cycles; and (c) the fatigue life S–N curves of experimental data and theoretical analysis for cross-ply SiC/CAS composite at room temperature.

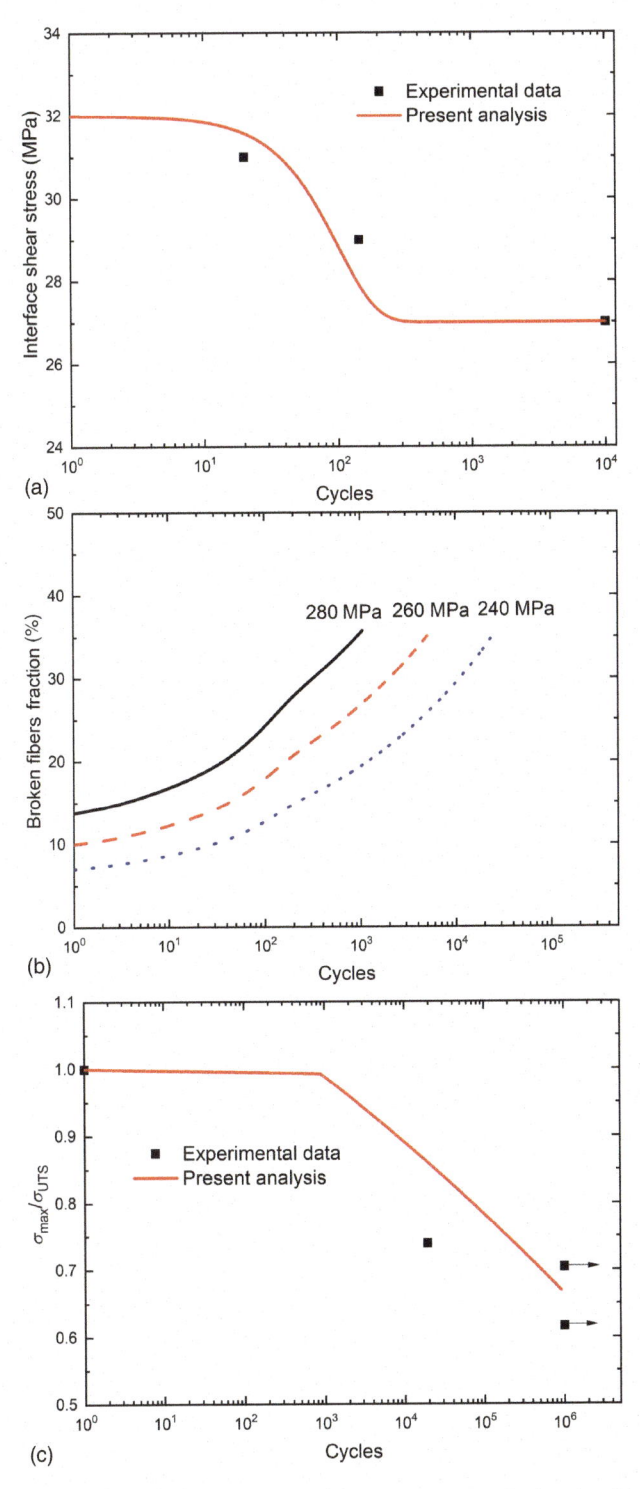

Figure 2.30 (a) The interface shear stress versus applied cycles; (b) the broken fibers fraction versus applied cycles; and (c) the fatigue life S–N curves of experimental data and theoretical analysis for cross-ply SiC/1723 composite at room temperature.

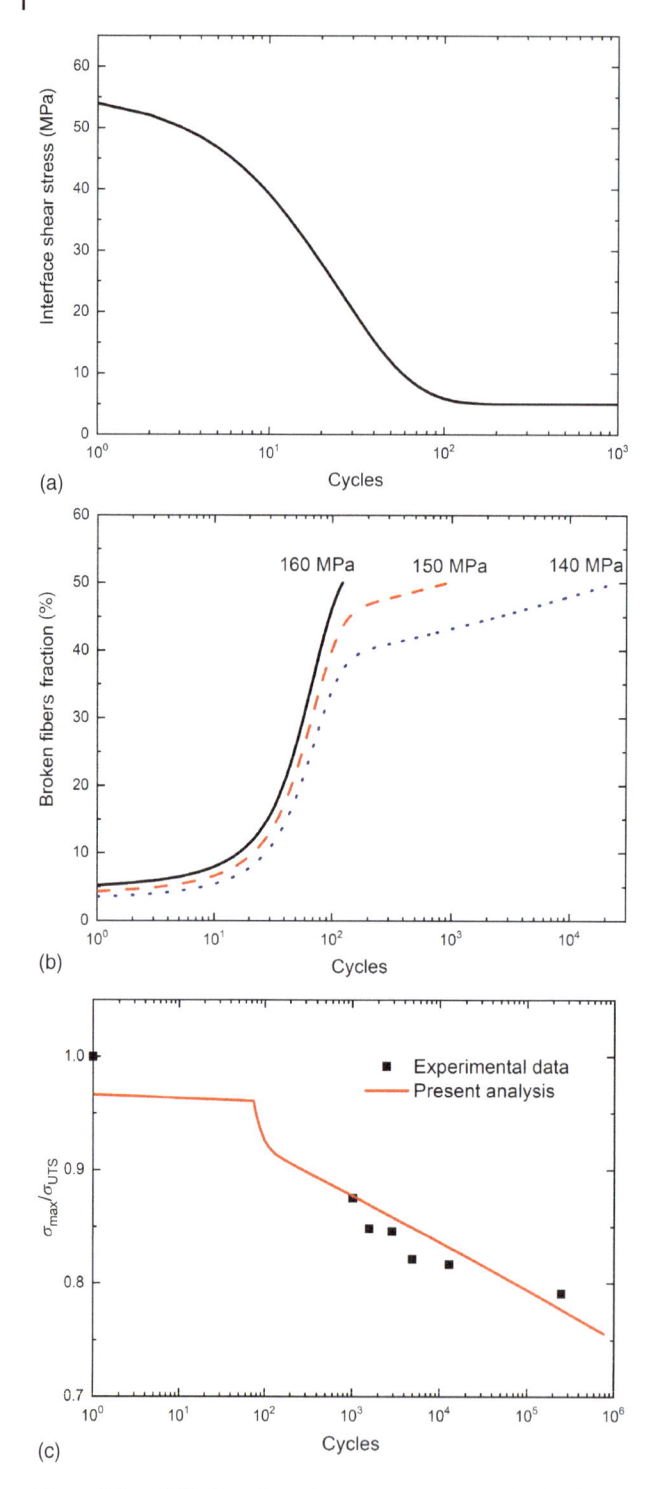

Figure 2.31 (a) The interface shear stress versus applied cycles; (b) the broken fibers fraction versus applied cycles; and (c) the fatigue life S–N curves of experimental data and theoretical analysis for 2D SiC/SiC composite at room temperature.

$V_f = 45\%$, $\Lambda = 0.5$, $r_f = 3.5\,\mu m$, $m_f = 5$, $p_1 = 0.018$, and $p_2 = 1.0$. The broken fibers fraction versus cycle number curves under the fatigue peak stresses of $\sigma_{max} = 400$, 380, and 360 MPa are shown in Figure 2.32b. Under the fatigue peak stress of $\sigma_{max} = 400$ MPa, the composite fatigue failed after 1487 cycles; under the fatigue peak stresses of $\sigma_{max} = 380$ and 360 MPa, the composite failed after 10 312 and 189 202 cycles, respectively. The experimental and theoretical fatigue life S–N curves are shown in Figure 2.32c, in which the fatigue limit approaches 85% of tensile strength.

For 2.5D C/SiC composite, the interface shear stress versus applied cycles curve is simulated by the Evans–Zok–McMeeking model (Evans et al. 1995), as shown in Figure 2.33a, in which the model parameters are given by $\tau_{io} = 20$ MPa, $\tau_{imin} = 8$ MPa, $\omega = 0.001$, and $\lambda = 1.0$. The material properties are given by $V_f = 40\%$, $\Lambda = 0.75$, $r_f = 3.5\,\mu m$, $m_f = 5$, $p_1 = 0.02$, and $p_2 = 1.2$. The broken fibers fraction versus applied cycle number curves under the fatigue peak stresses of $\sigma_{max} = 200$ and 180 MPa are shown in Figure 2.33b. Under the fatigue peak stress of $\sigma_{max} = 200$ MPa, the composite fatigue failed after 832 cycles; and under the fatigue peak stress of $\sigma_{max} = 180$ MPa, the composite fatigue failed after 13 470 cycles. The experimental and theoretical fatigue life S–N curves are shown in Figure 2.33c, in which the fatigue limit stress approaches 70% of tensile strength.

For 3D C/SiC composite, the interface shear stress versus applied cycles curve is simulated by Evans–Zok–McMeeking model (Evans et al. 1995), as shown in Figure 2.34a, in which the model parameters are given by $\tau_{io} = 20$ MPa, $\tau_{imin} = 5$ MPa, $\omega = 0.02$, and $\lambda = 1.0$. The material properties are given by $V_f = 40\%$, $\Lambda = 0.93$, $r_f = 3.5\,\mu m$, $m_f = 5$, $p_1 = 0.012$, and $p_2 = 1.0$. The broken fibers fraction versus cycle number curves under the fatigue peak stresses of $\sigma_{max} = 270$ and 250 MPa are shown in Figure 2.34b. Under the fatigue peak stress of $\sigma_{max} = 270$ MPa, the composite fatigue failed after 135 cycles; and under the fatigue peak stress of $\sigma_{max} = 250$ MPa, the composite fatigue failed after 9754 cycles. The experimental and theoretical fatigue life S–N curves are shown in Figure 2.34c, in which the fatigue limit approaches 85% of tensile strength.

2.4.2.2 Life Prediction at Elevated Temperature

The tensile strength of unidirectional C/SiC composite is $\sigma_{UTS} = 320$ MPa at 800 °C in air condition, and the fatigue peak stresses are 0.37, 0.43, 0.56, 0.65, and 0.78 of tensile strength; the tensile strength of cross-ply C/SiC composite is $\sigma_{UTS} = 150$ MPa at 800 °C in air condition, and the fatigue peak stresses are 0.60, 0.65, and 0.70 of tensile strength; the tensile strength of 2D C/SiC composite is $\sigma_{UTS} = 487$ MPa at 550 °C in air condition, and the fatigue peak stresses are 0.22, 0.36, 0.56, and 0.72 of tensile strength (Mall and Engesser 2006); the tensile strength of 2D C/SiC composite is $\sigma_{UTS} = 300$ MPa at 1300 °C in the oxidative environment, and the fatigue peak stresses are 0.5, 0.6, 0.7, and 0.8 of tensile strength (Cheng et al. 2010); the tensile strength of 2.5D C/SiC composite is $\sigma_{UTS} = 280$ MPa at 800 °C in air, and the fatigue peak stresses are 0.5, 0.6, 0.7, and 0.8 of tensile strength (Yang 2011); the tensile strength of 2.5D C/SiC composite is $\sigma_{UTS} = 228$ MPa at 900 °C in air condition, and the fatigue peak stresses are 0.35, 0.4, 0.43, 0.52, 0.6, and 0.7 of tensile strength (Zhang et al. 2013) and the tensile strength of 3D C/SiC composite is $\sigma_{UTS} = 304$ MPa at 1300 °C in vacuum

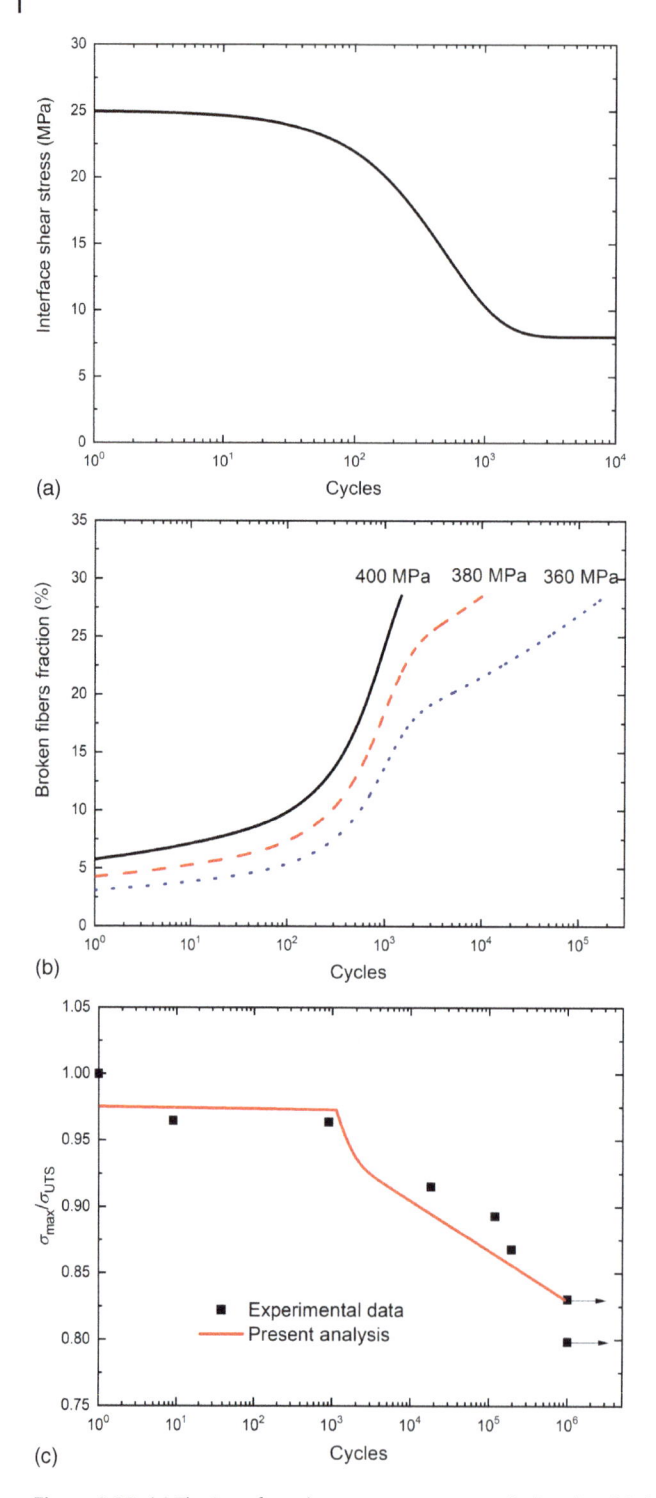

Figure 2.32 (a) The interface shear stress versus applied cycles; (b) the broken fibers fraction versus applied cycles; and (c) the fatigue life S–N curves of experimental data and theoretical analysis for 2D C/SiC composite at room temperature.

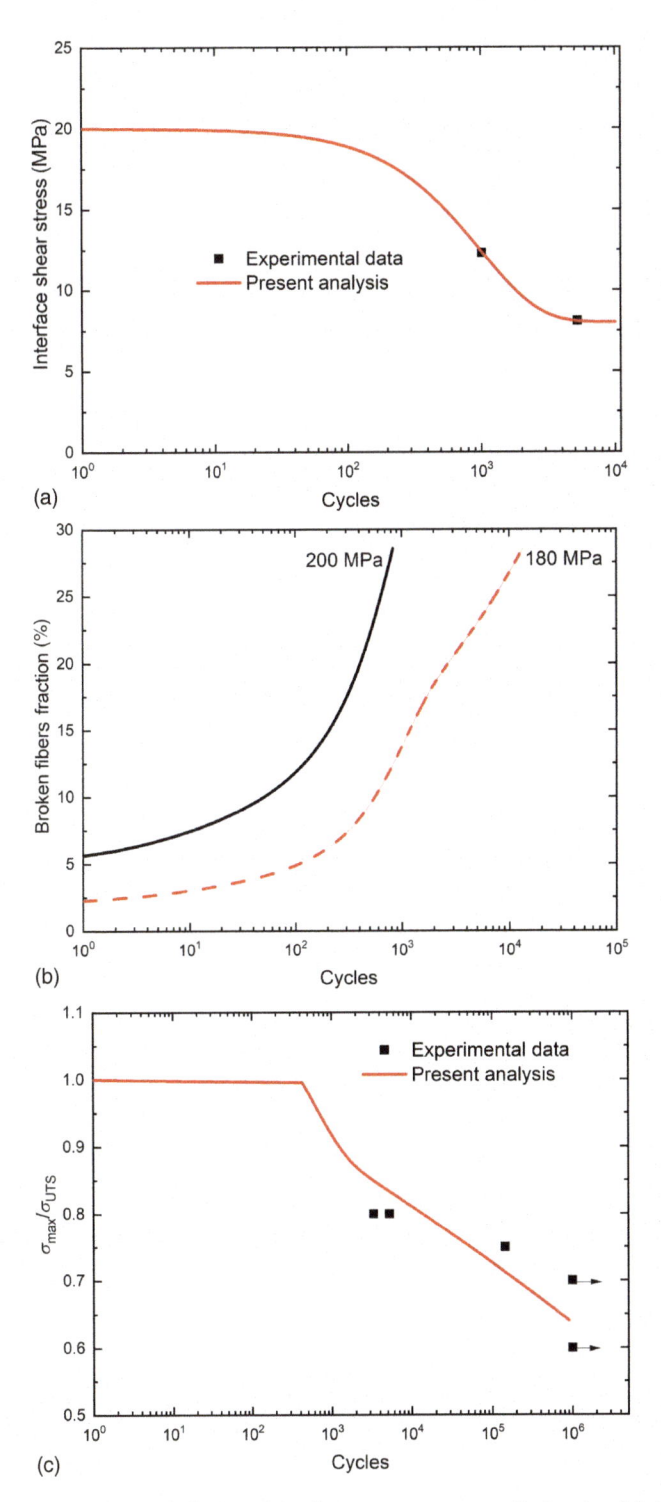

Figure 2.33 (a) The interface shear stress versus applied cycles; (b) the broken fibers fraction versus applied cycles; and (c) the fatigue life S–N curves of experimental data and theoretical analysis for 2.5D C/SiC composite at room temperature.

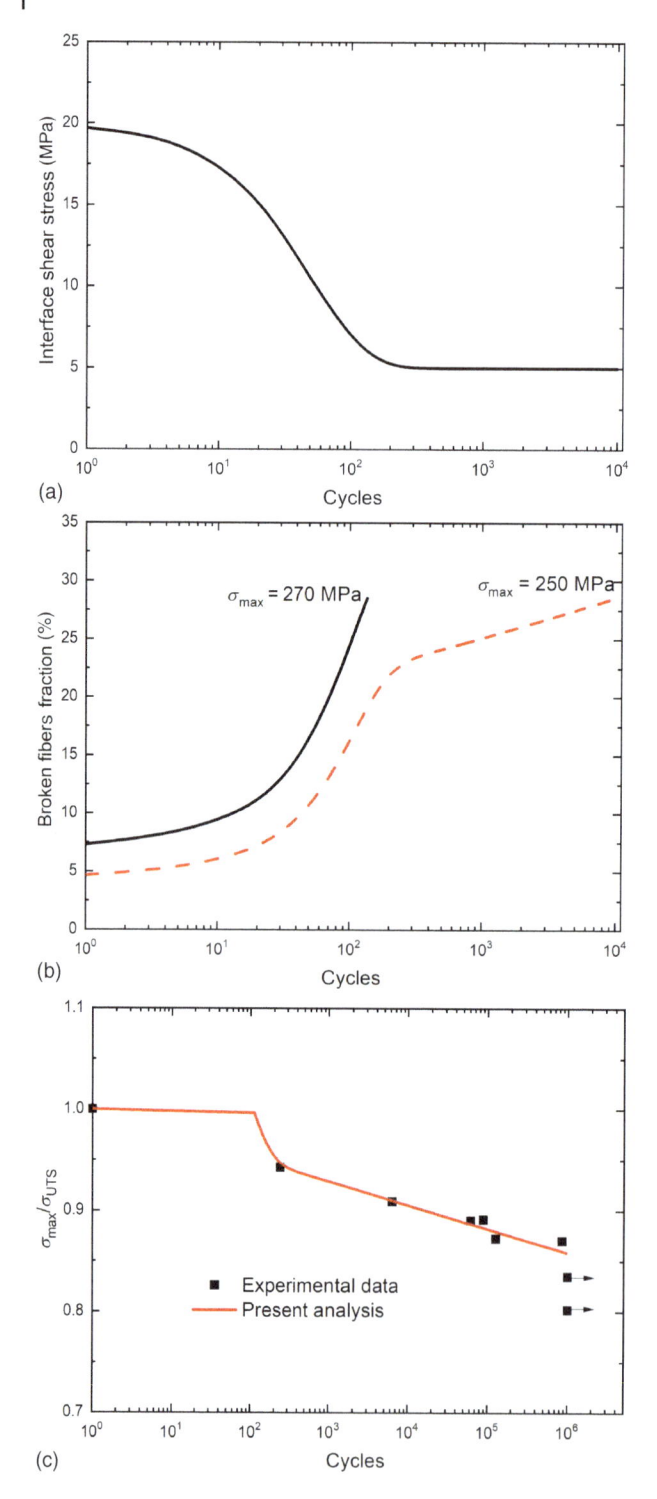

Figure 2.34 (a) The interface shear stress versus applied cycles; (b) the broken fibers fraction versus applied cycles; and (c) the fatigue life S–N curves of experimental data and theoretical analysis for 3D C/SiC composite at room temperature.

condition, and the fatigue peak stresses are 0.83, 0.5, 0.93, 0.98, and 0.99 of tensile strength (Du et al. 2002).

For unidirectional C/SiC composite at 800 °C in air, the interface shear stress versus applied cycles curve is simulated by the Evans–Zok–McMeeking model (Evans et al. 1995), as shown in Figure 2.35a, in which the model parameters are given by $\tau_{io} = 6.1$ MPa, $\tau_{imin} = 0.2$ MPa, $\omega = 0.001$, and $\lambda = 0.8$. The material properties are given by $V_f = 40\%$, $\Lambda = 1.0$, $r_f = 3.5$ μm, $k = \exp[11.383 - (8716/T_{em})] \times 10^{-18}/60$ m^2/s, $K_{IC} = 0.5$ MPa/m$^{1/2}$, $Y = 1$, $\varphi_1 = 7.021 \times 10^{-3} \times \exp(8231/T_{em})$, $\varphi_2 = 227.1 \times \exp(-17\,090/T_{em})$, $m_f = 5$, $l_0 = 25 \times 10^{-3}$ m, $p_1 = 0.02$, and $p_2 = 1.0$. The broken fibers fraction versus cycle number curves under the fatigue peak stresses of $\sigma_{max} = 240$ and 200 MPa are shown in Figure 2.35b. Under the fatigue peak stress of $\sigma_{max} = 240$ MPa, the composite fatigue failed after 2970 cycles with the broken fibers fraction of 20.4%; and under the fatigue peak stress of $\sigma_{max} = 200$ MPa, the composite fatigue failed after 9195 cycles with the broken fibers fraction of $q = 15.3\%$. The experimental and theoretical fatigue life S–N curves are shown in Figure 2.35c, in which the fatigue life at 800 °C in air condition is greatly reduced compared with that at room temperature, mainly attributed to oxidation of pyrolytic carbon (PyC) interphase and carbon fibers.

For cross-ply C/SiC composite at 800 °C in air condition, the interface shear stress versus applied cycles curve is simulated by the Evans–Zok–McMeeking model (Evans et al. 1995), as shown in Figure 2.36a, in which the model parameters are given by $\tau_{io} = 5.5$ MPa, $\tau_{imin} = 0.4$ MPa, $\omega = 0.001$, and $\lambda = 1.0$. The material properties are given by: $V_f = 40\%$, $\Lambda = 0.5$, $r_f = 3.5$ μm, $k = \exp[11.383 - (8716/T_{em})] \times 10^{-18}/60$ m^2/s, $K_{IC} = 0.5$ MPa/m$^{1/2}$, $Y = 1$, $\varphi_1 = 7.021 \times 10^{-3} \times \exp(8231/T_{em})$, $\varphi_2 = 227.1 \times \exp(-17\,090/T_{em})$, $m_f = 5$, $l_0 = 25 \times 10^{-3}$ m, $p_1 = 0.02$, and $p_2 = 1.0$. The broken fibers fraction versus applied cycle number curves under the fatigue peak stresses of $\sigma_{max} = 100$ and 80 MPa are shown in Figure 2.36b. Under the fatigue peak stress of $\sigma_{max} = 100$ MPa, the composite fatigue failed after 2134 cycles with the broken fibers fraction of $q = 18.5\%$; and under the fatigue peak stress of $\sigma_{max} = 80$ MPa, the composite fatigue failed after 9881 cycles with the broken fibers fraction of $q = 26.8\%$. The experimental and theoretical fatigue life S–N curves are shown in Figure 2.36c, in which the fatigue life at 800 °C in air condition is greatly reduced compared with that at room temperature, attributed to oxidation of PyC interphase and carbon fibers.

For 2D C/SiC composite at 550 °C in air condition, the interface shear stress versus applied cycles is simulated by the Evans–Zok–McMeeking model (Evans et al. 1995), as shown in Figure 2.37a, in which the model parameters are given by $\tau_{io} = 25$ MPa, $\tau_{imin} = 8$ MPa, $\omega = 0.0001$, and $\lambda = 1.2$. The material properties are given by $V_f = 45\%$, $\Lambda = 0.5$, $r_f = 3.5$ μm, $k = \exp[11.383 - (8716/T_{em})] \times 10^{-18}/60$ m^2/s, $K_{IC} = 0.5$ MPa/m$^{1/2}$, $Y = 1$, $\varphi_1 = 7.021 \times 10^{-3} \times \exp(8231/T_{em})$, $\varphi_2 = 227.1 \times \exp(-17\,090/T_{em})$, $m_f = 5$, $l_0 = 25 \times 10^{-3}$ m, $p_1 = 0.02$, and $p_2 = 1.0$. The broken fibers fraction versus applied cycle number curves under the fatigue peak stresses of $\sigma_{max} = 420$ and 320 MPa are shown in Figure 2.37b. Under the fatigue peak stress of $\sigma_{max} = 420$ MPa, the composite fatigue failed after 25 cycles with the broken fibers fraction of

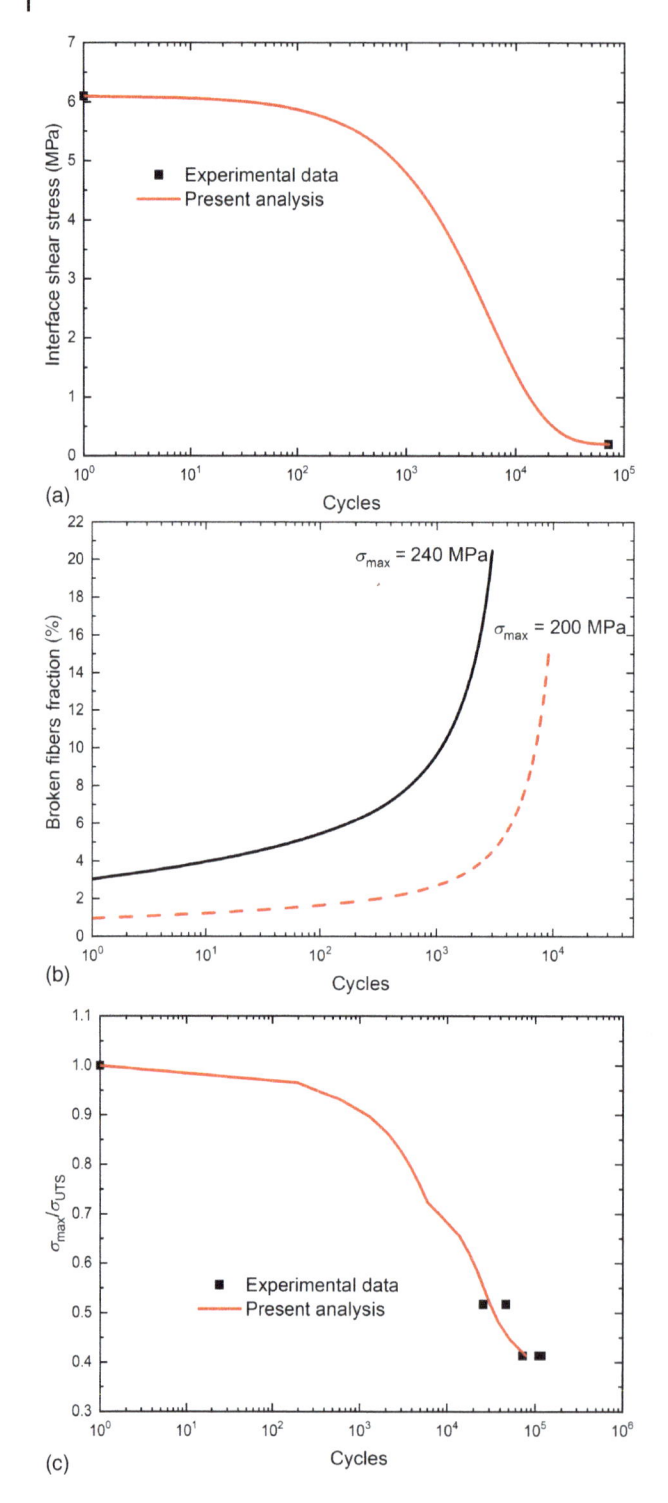

Figure 2.35 (a) The interface shear stress versus applied cycles; (b) the broken fibers fraction versus applied cycles; and (c) the fatigue life S–N curves of experimental data and theoretical analysis for unidirectional C/SiC composite at 800 °C in air condition.

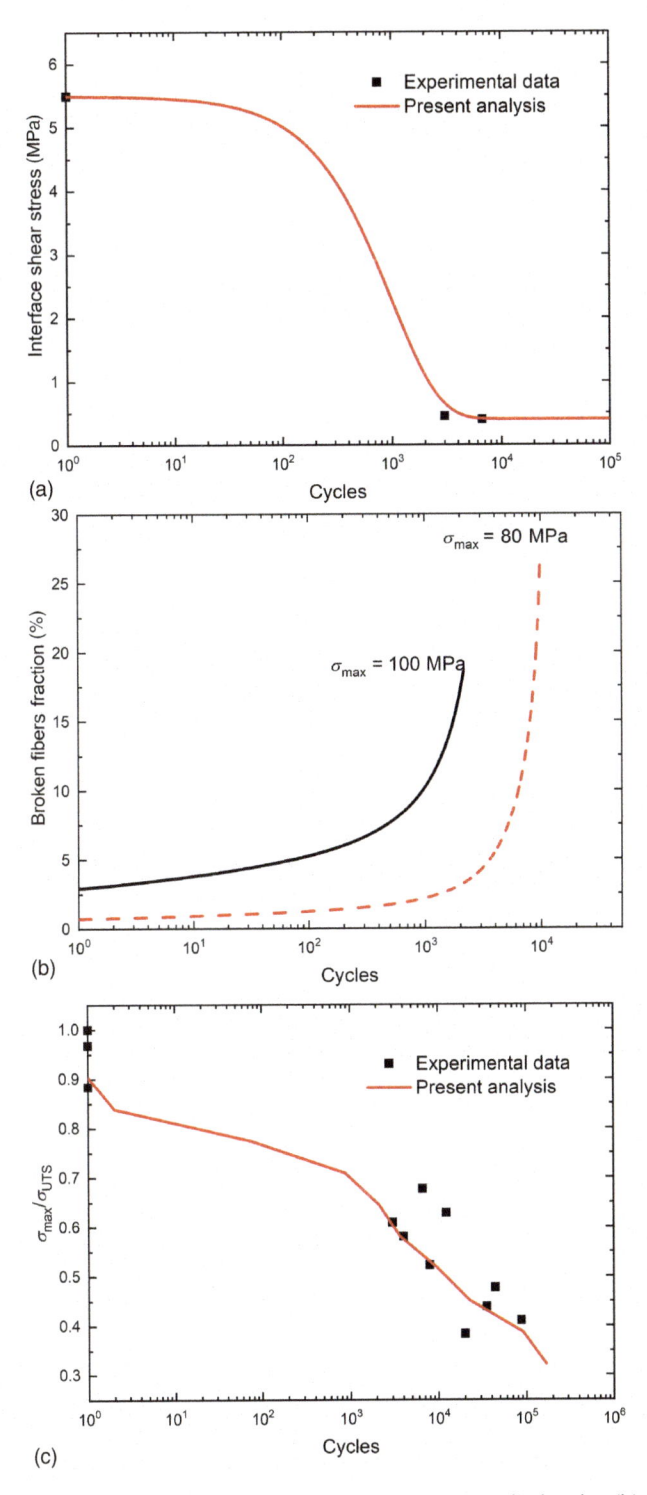

Figure 2.36 (a) The interface shear stress versus applied cycles; (b) the broken fibers fraction versus applied cycles; and (c) the fatigue life S–N curves of experimental data and theoretical analysis for cross-ply C/SiC composite at 800 °C in air condition.

$q = 27.8\%$; and under the fatigue peak stress of $\sigma_{max} = 320$ MPa, the composite fatigue failed after 12 457 cycles with the broken fibers fraction of $q = 26.2\%$. The experimental and theoretical fatigue life S–N curves are shown in Figure 2.37c, in which the fatigue life at 550 °C in air condition is greatly reduced compared with that at room temperature, mainly attributed to oxidation of interphase and carbon fibers.

For 2D C/SiC composite at 1300 °C in the oxidative environment, the interface shear stress versus applied cycles is simulated by the Evans–Zok–McMeeking model (Evans et al. 1995), as shown in Figure 2.38a, in which the model parameters are given by $\tau_{io} = 20$ MPa, $\tau_{imin} = 6$ MPa, $\omega = 0.0005$, and $\lambda = 1.2$. The material properties are given by $V_f = 40\%$, $\Lambda = 0.5$, $r_f = 3.5$ μm, $k = \exp[11.383 - (8716/T_{em})] \times 10^{-18}/60$ m^2/s, $K_{IC} = 0.5$ MPa/m$^{1/2}$, $Y = 1$, $\varphi_1 = 7.021 \times 10^{-3} \times \exp(8231/T_{em})$, $\varphi_2 = 227.1 \times \exp(-17\,090/T_{em})$, $m_f = 5$, $l_0 = 25 \times 10^{-3}$ m, $p_1 = 0.02$, and $p_2 = 1.0$. The broken fibers fraction versus applied cycle number curves under the fatigue peak stresses of $\sigma_{max} = 250$ and 200 MPa are shown in Figure 2.38b. Under the fatigue peak stress of $\sigma_{max} = 250$ MPa, the composite fatigue failed after 196 cycles with the broken fibers fraction of $q = 28.4\%$; and under the fatigue peak stress of $\sigma_{max} = 200$ MPa, the composite fatigue failed after 11 480 cycles with the broken fibers fraction of $q = 23.7\%$. The experimental and theoretical fatigue life S–N curves are shown in Figure 2.38c, in which the fatigue life at 1300 °C in the oxidative environment is greatly reduced compared with that at room temperature, mainly attributed to oxidation of interphase and carbon fibers.

For 2.5D C/SiC composite at 800 °C in air condition, the interface shear stress versus applied cycles is simulated by the Evans–Zok–McMeeking model (Evans et al. 1995), as shown in Figure 2.39a, in which the model parameters are given by $\tau_{io} = 20$ MPa, $\tau_{imin} = 5$ MPa, $\omega = 0.008$, and $\lambda = 1.2$. The material properties are given by $V_f = 40\%$, $\Lambda = 0.75$, $r_f = 3.5$ μm, $k = \exp[11.383 - (8716/T_{em})] \times 10^{-18}/60$ m^2/s, $K_{IC} = 0.5$ MPa/m$^{1/2}$, $Y = 1$, $\varphi_1 = 7.021 \times 10^{-3} \times \exp(8231/T_{em})$, $\varphi_2 = 227.1 \times \exp(-17\,090/T_{em})$, $m_f = 5$, $l_0 = 25 \times 10^{-3}$ m, $p_1 = 0.03$, and $p_2 = 1.2$. The broken fibers fraction versus cycle number curves under the fatigue peak stresses of $\sigma_{max} = 200$ and 180 MPa are shown in Figure 2.39b. Under the fatigue peak stress of $\sigma_{max} = 200$ MPa, the composite fatigue failed after 2945 cycles with the broken fibers fraction of $q = 28.5\%$; and under the peak stress of $\sigma_{max} = 180$ MPa, the composite fatigue failed after 6078 cycles with the broken fibers fraction of $q = 20.1\%$. The experimental and predicted fatigue life S–N curves are shown in Figure 2.39c, in which the fatigue life at 800 °C in air is greatly reduced compared with that at room temperature, mainly attributed to oxidation of PyC interphase and carbon fibers.

For 2.5D C/SiC composite at 900 °C in air condition, the interface shear stress versus applied cycles is simulated by the Evans–Zok–McMeeking model (Evans et al. 1995), as shown in Figure 2.40a, in which the model parameters are given by $\tau_{io} = 25$ MPa, $\tau_{imin} = 5$ MPa, $\omega = 0.003$, and $\lambda = 1.2$. The material properties are given by $V_f = 40\%$, $\Lambda = 0.75$, $r_f = 3.5$ μm, $k = \exp[11.383 - (8716/T_{em})] \times 10^{-18}/60$ m^2/s, $K_{IC} = 0.5$ MPa/m$^{1/2}$, $Y = 1$, $\varphi_1 = 7.021 \times 10^{-3} \times \exp(8231/T_{em})$, $\varphi_2 = 227.1 \times \exp(-17\,090/T_{em})$, $m_f = 5$, $l_0 = 25 \times 10^{-3}$ m, $p_1 = 0.03$, and $p_2 = 1.2$. The broken fibers fraction versus

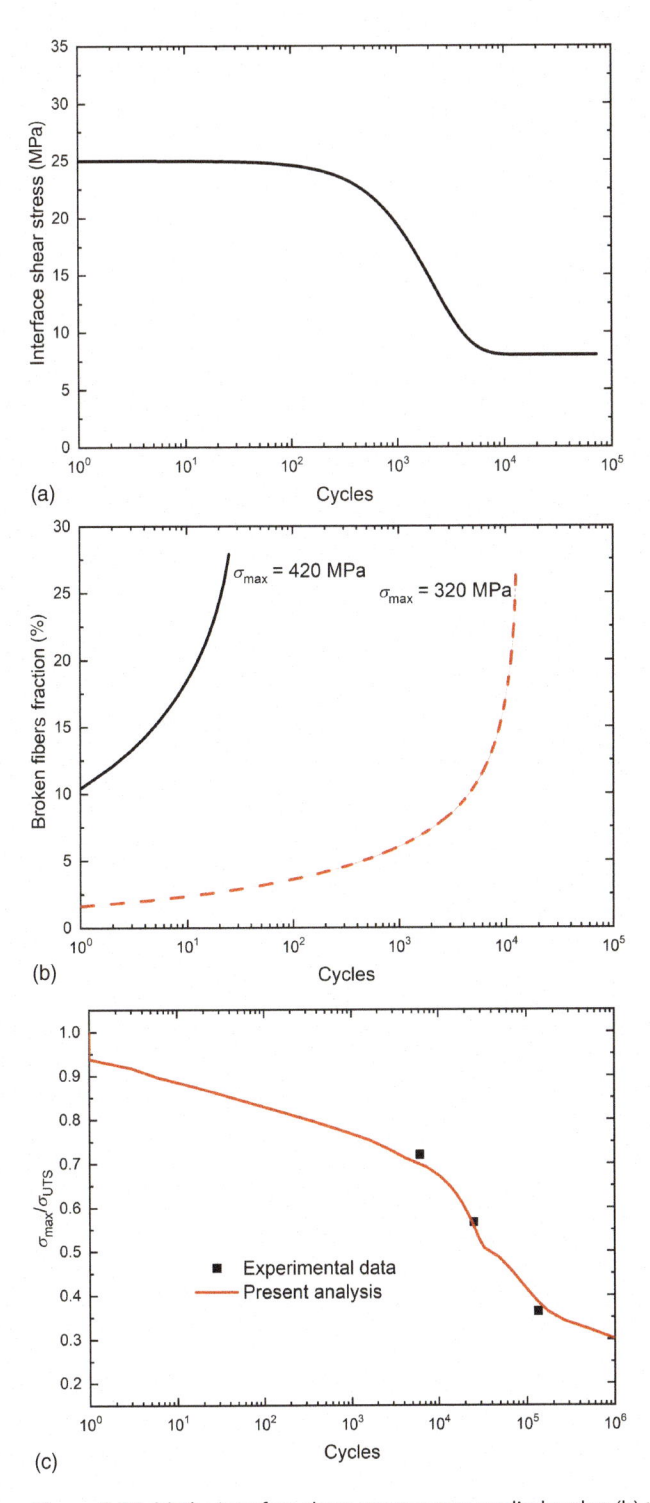

2.4 Effect of Interface Properties on Lifetime of Ceramic-Matrix Composites | 97

Figure 2.37 (a) The interface shear stress versus applied cycles; (b) the broken fibers fraction versus applied cycles; and (c) the fatigue life S–N curves of experimental data and theoretical analysis for 2D C/SiC composite at 550 °C in air condition.

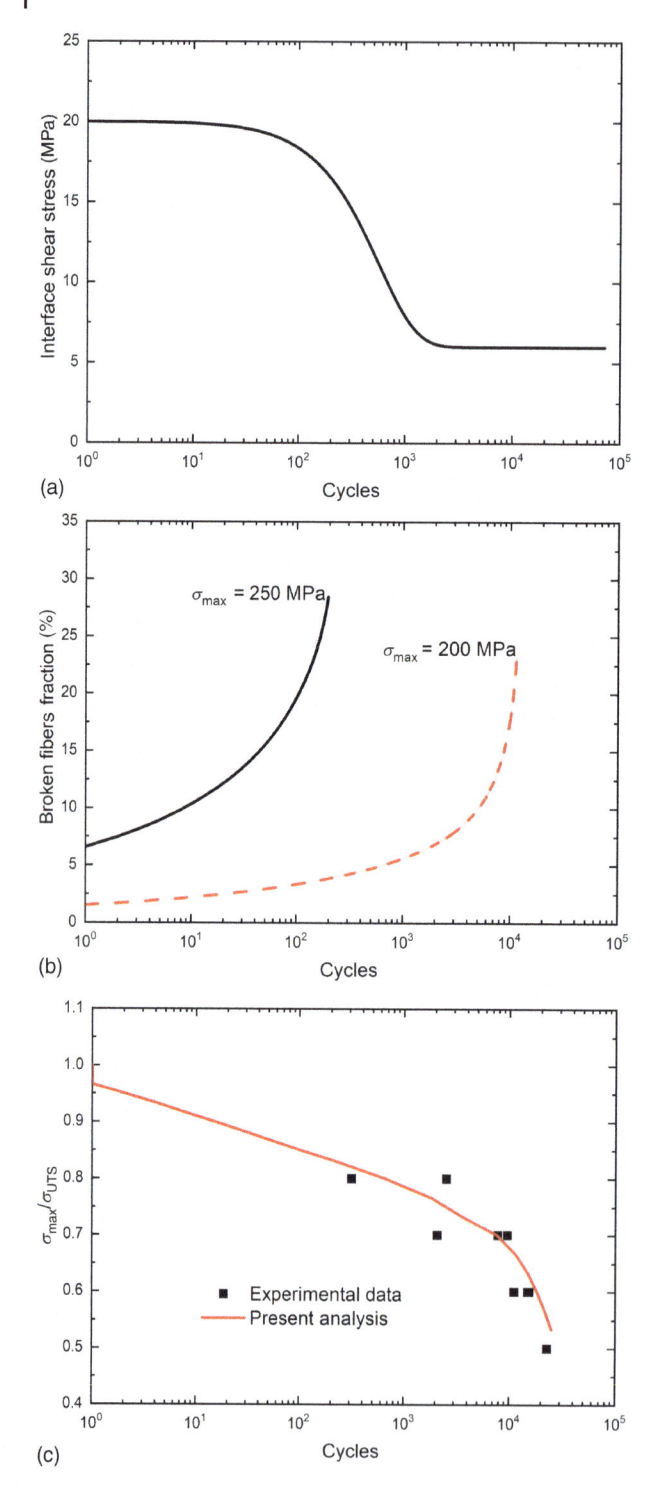

Figure 2.38 (a) The interface shear stress versus applied cycles; (b) the broken fibers fraction versus applied cycles; and (c) the fatigue life S–N curves of experimental data and theoretical analysis for 2D C/SiC composite at 1300 °C in the oxidative atmosphere.

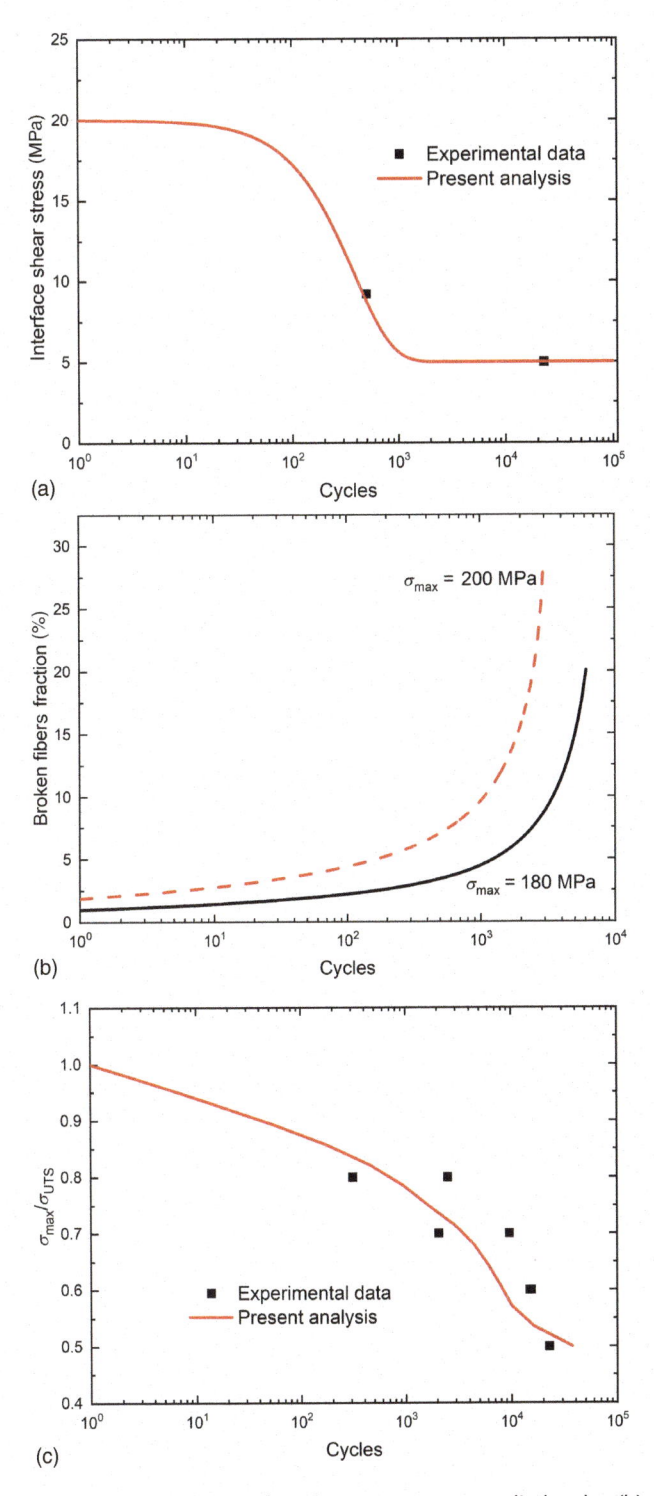

Figure 2.39 (a) The interface shear stress versus applied cycles; (b) the broken fibers fraction versus applied cycles; and (c) the fatigue life S–N curves of experimental data and theoretical analysis for 2.5D C/SiC composite at 800 °C in air condition.

cycle number curves under the fatigue peak stresses of σ_{max} = 220 and 210 MPa are shown in Figure 2.40b. Under the fatigue peak stress of σ_{max} = 220 MPa, the composite fatigue failed after 2945 cycles with the broken fibers fraction of q = 26.4%; and under the peak stress of σ_{max} = 210 MPa, the composite fatigue failed after 1649 cycles with the broken fibers fraction of q = 24.1%. The experimental and theoretical fatigue life S–N curves are shown in Figure 2.40c, in which the fatigue life at 900 °C in air is greatly reduced compared with that at room temperature, mainly attributed to oxidation of PyC interphase and carbon fibers.

For 3D C/SiC composite at 1300 °C in vacuum condition, the interface shear stress versus applied cycles is simulated by the Evans–Zok–McMeeking model (Evans et al. 1995), as shown in Figure 2.41a, in which the model parameters are given by τ_{io} = 20 MPa, τ_{imin} = 8 MPa, ω = 0.002, and λ = 1.0. The material properties are given by V_f = 40%, Λ = 0.93, r_f = 3.5 μm, p_1 = 0.018, and p_2 = 1.0. The broken fibers fraction versus cycle number curves under the fatigue peak stresses of σ_{max} = 300 and 295 MPa are shown in Figure 2.41b. Under the peak stress of σ_{max} = 300 MPa, the composite fatigue failed after 14 968 cycles with the broken fibers fraction of q = 28.5%; and under the peak stress of σ_{max} = 295 MPa, the composite fatigue failed after 25 773 cycles with the broken fibers fraction of q = 28.5%. The experimental and predicted fatigue life S–N curves at 1300 °C in vacuum condition are shown in Figure 2.41c, in which the fatigue life and fatigue limit at 1300 °C in vacuum condition are increased compared with those at room temperature.

2.5 Conclusion

In this chapter, the effect of the fiber/matrix interface properties and pre-exposure on the tensile and fatigue behavior of fiber-reinforced CMCs is investigated. The relationships between the interface properties and the composite tensile and fatigue damage are established. The effects of the interface properties and pre-exposure temperature and time on the first matrix cracking stress, matrix cracking evolution, first and complete interface debonding stress, fatigue hysteresis dissipated energy, fatigue hysteresis modulus, and fatigue hysteresis width, and tensile damage and fracture processes are analyzed. The fatigue life of fiber-reinforced CMCs with different fiber preforms, i.e. unidirectional, cross-ply, 2D, 2.5D, and 3D CMCs at room and elevated temperatures in air and oxidative environment, are predicted using the micromechanics approach. The experimental tensile and fatigue damage and fatigue life of different CMCs are predicted for different interface properties and testing conditions.

(1) When the interface shear stress increases, the first matrix cracking stress increases and the broken fibers fraction increases; the matrix cracking density, the saturation matrix cracking stress, and the interface debonding length increase; the initial interface debonding stress and the complete interface debonding stress increase, the hysteresis dissipated energy decreases, the hysteresis modulus increases, and the hysteresis width decreases.

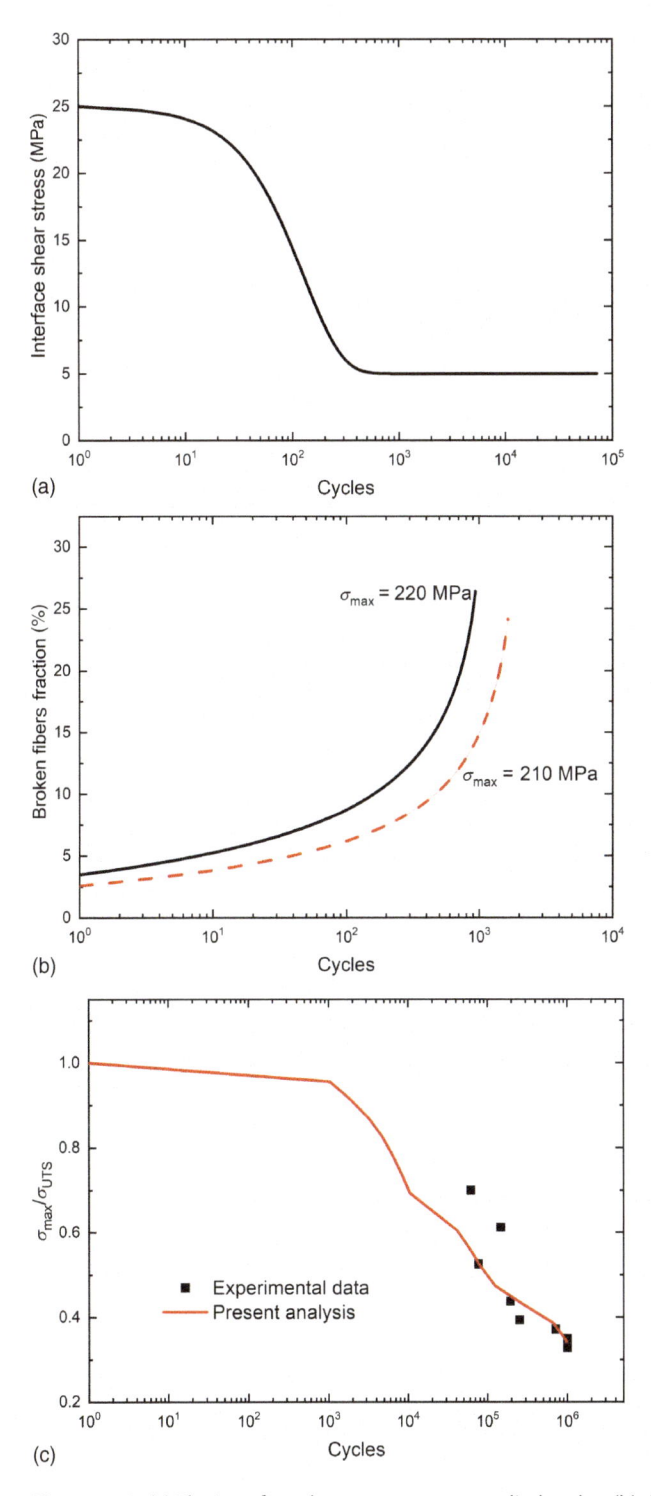

Figure 2.40 (a) The interface shear stress versus applied cycles; (b) the broken fibers fraction versus applied cycles; and (c) the fatigue life S–N curves of experimental data and theoretical analysis for 2.5D C/SiC composite at 900 °C in air condition.

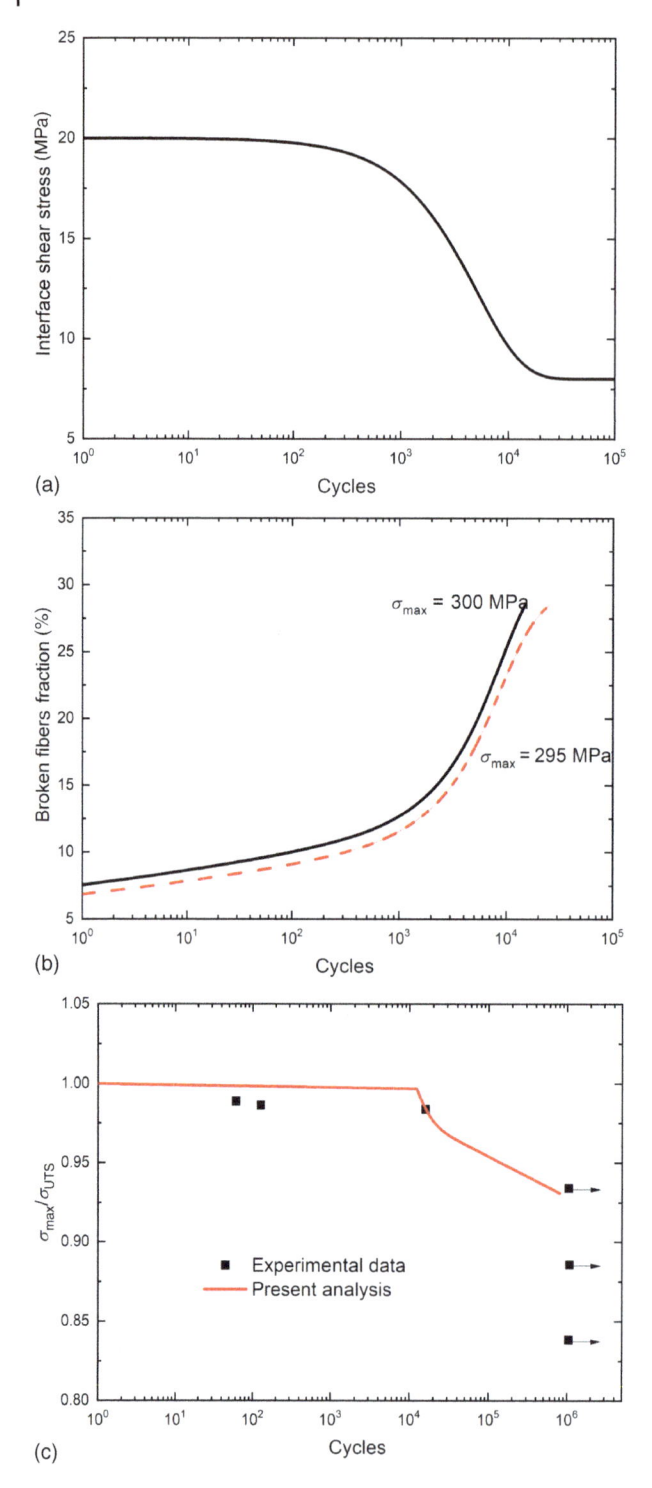

Figure 2.41 (a) The interface shear stress versus applied cycles; (b) the broken fibers fraction versus applied cycles; and (c) the fatigue life S–N curves of experimental data and theoretical analysis for 3D C/SiC composite at 1300 °C in vacuum condition.

(2) When the interface debonding energy increases, the first matrix cracking stress increases and the broken fibers fraction increases; the matrix cracking density decreases and the saturation matrix cracking stress increases; and the initial interface debonding stress and the complete interface debonding stress increase, the hysteresis dissipated energy decreases, the hysteresis modulus increases, and the hysteresis width decreases.

(3) When the pre-exposure temperature and time increase, the composite tensile strength and failure strain both decrease; the interface debonded length and the interface oxidation ratio both increase; and the fiber broken fraction increases at low applied stress level.

(4) The broken fibers fraction versus applied cycles curve can be divided into two regions, i.e. at the initial loading cycles, the broken fibers fraction increases rapidly due to the degradation of interface shear stress and fibers strength; and when interface shear stress approaches to the steady-state value, fiber failure is mainly controlled by fibers strength degradation, which makes the broken fibers fraction increase slowly.

(5) The predicted fatigue life S–N curves can also be divided into two regions, i.e. the region I is controlled by the degradation of interface shear stress and fibers strength; and the region II is only controlled by the degradation of fibers strength.

(6) The fatigue life of unidirectional, cross-ply, 2D, and 2.5D C/SiC composites at elevated temperatures in air or oxidative environment is greatly reduced compared with that at room temperature, mainly attributed to oxidation of PyC interphase and carbon fibers; however, at 1300 °C in vacuum, the fatigue life and fatigue limit increase compared with those at room temperature.

References

Carrere, N., Martin, E., and Lamon, J. (2000). The influence of the interphase and associated interfaces on the deflection of matrix cracks in ceramic matrix composites. *Composites Part A Applied Science and Manufacturing* 31: 1179–1190. https://doi.org/10.1016/S1359-835X(00)00095-6.

Casas, L. and Martinez-Esnaola, J.M. (2003). Modeling the effect of oxidation on the creep behavior of fiber-reinforced ceramic matrix composites. *Acta Materialia* 51: 3745–3757. https://doi.org/10.1016/S1359-6454(03)00189-7.

Cheng, Q.Y., Tong, X.Y., Zheng, X. et al. (2010). Experimental investigation on the fatigue characteristics about high temperature of plain-woven C/SiC composite. *Journal of Mechanical Strength* 32: 819–825.

Cho, C.D., Holmes, J.W., and Barber, J.R. (1991). Estimation of interfacial shear in ceramic composites from frictional heating measurements. *Journal of the American Ceramic Society* 74: 2802–2808. https://doi.org/10.1111/j.1151-2916.1991.tb06846.x.

Curtin, W.A. (1991). Theory of mechanical properties of ceramic-matrix composites. *Journal of the American Ceramic Society* 74: 2837–2845. https://doi.org/10.1111/j.1151-2916.1991.tb06852.x.

Curtin, W.A. (1993). Multiple matrix cracking in brittle matrix composites. *Acta Metallurgica et Materialia* 41: 1369–1377. https://doi.org/10.1016/0956-7151(93)90246-O.

Curtin, W.A. (2000). Stress-strain behavior of brittle matrix composites. In: *Comprehensive Composite Materials*, vol. 4, 47–76. Elsevier Science Ltd.. https://doi.org/10.1016/B0-08-042993-9/00088-7.

Curtin, W.A., Ahn, B.K., and Takeda, N. (1998). Modeling brittle and tough stress-strain behavior in unidirectional ceramic matrix composites. *Acta Materialia* 46: 3409–3420. https://doi.org/10.1016/S1359-6454(98)00041-X.

Dassios, K.G., Aggelis, D.G., Kordatos, E.Z., and Matikas, T.E. (2013). Cyclic loading of a SiC-fiber reinforced ceramic matrix composite reveals damage mechanisms and thermal residual stress state. *Composites Part A Applied Science and Manufacturing* 44: 105–113. https://doi.org/10.1016/j.compositesa.2012.06.011.

DiCarlo, J.A. and Roode, M. (2006). Ceramic composite development for gas turbine hot section components. *ASME Turbo Expo 2006: Power for Land, Sea, and Air*, Barcelona, Spain (8–11 May 2006). https://doi.org/10.1115/GT2006-90151.

DiCarlo, J.A., Yun, H.M., Morscher, G.N., and Bhatt, R.T. (2005). SiC/SiC composites for 1200 °C and above. In: *Handbook of Ceramic Composites* (ed. N.P. Bansal). Boston, MA: Springer. https://doi.org/10.1007/0-387-23986-3_4.

Ding, D. (2014). Processing, properties and applications of ceramic matrix composites, SiC/SiC: an overview. In: *Advances in Ceramic Matrix Composites* (ed. I.M. Low). Woodhead Publishing. ISBN: 978-0-85709-120-8. https://doi.org/10.1016/B978-0-08-102166-8.00002-5.

Domergue, J.M., Vagaggini, E., and Evans, A.G. (1995). Relationship between hysteresis measurements and the constituent properties of ceramic matrix composites: II. Experimental studies on unidirectional materials. *Journal of the American Ceramic Society* 78: 2721–2731. https://doi.org/10.1111/j.1151-2916.1995.tb08047.x.

Du, S.M., Qiao, S.R., Ji, G.C., and Han, D. (2002). Tension-tension fatigue behavior of 3D-C/SiC composite at room temperature and 1300 °C. *Materials Engineering* 9: 22–25.

Evans, A.G., Zok, F.W., and McMeeking, R.M. (1995). Fatigue of ceramic matrix composites. *Acta Metallurgica et Materialia* 43: 859–875. https://doi.org/10.1016/0956-7151(94)00304-Z.

Fantozzi, G. and Reynaud, P. (2009). Mechanical hysteresis in ceramic matrix composites. *Materials Science and Engineering A* 521–522: 18–23. https://doi.org/10.1016/j.msea.2008.09.128.

Gao, Y., Mai, Y., and Cotterell, B. (1988, 1988). Fracture of fiber-reinforced materials. *Journal of Applied Mathematics and Physics* 39: 550–572. https://doi.org/10.1007/BF00948962.

Halbig, M.C., Jaskowiak, M.H., Kiser, J.D., and Zhu, D. (2013). Evaluation of Ceramic Matrix Composite Technology for Aircraft Turbine Engine Applications. NASA report. https://ntrs.nasa.gov/archive/nasa/casi.ntrs.nasa.gov/20130010774.pdf.

Han, Z. and Morscher, G.N. (2015). Electrical resistance and acoustic emission during fatigue testing of SiC/SiC composites. In: *Mechanical Properties and Performance of Engineering Ceramics and Composites X: A Collection of Papers*

Presented at the 39th International Conference on Advanced Ceramics and Composites. https://doi.org/10.1002/9781119211310.ch4.

Holmes, J.W. and Cho, C.D. (1992). Experimental observation of frictional heating in fiber-reinforced ceramics. *Journal of the American Ceramic Society* 75: 929–938. https://doi.org/10.1111/j.1151-2916.1992.tb04162.x.

Kumar, R.S., Mordasky, M., and Ojard, G. (2018). Delamination fracture in ceramic matrix composites: from coupons to components. *ASME Turbo Expo 2018: Turbomachinery Technical Conference and Exposition*, Oslo, Norway (11–15 June 2018). https://doi.org/10.1115/GT2018-75571.

Lara-Curzio, E. (1999). Analysis of oxidation-assisted stress-rupture of continuous fiber-reinforced ceramic matrix composites at intermediate temperatures. *Composites Part A Applied Science and Manufacturing* 30: 549–554. https://doi.org/10.1016/S1359-835X(98)00148-1.

Li, L. (2016). Fatigue life prediction of fiber-reinforced ceramic-matrix composites with different fiber preforms at room and elevated temperatures. *Materials* 9: 207. https://doi.org/10.3390/ma9030207.

Li, L. (2018a). *Damage, Fracture and Fatigue of Ceramic-Matrix Composites*. Springer Nature Singapore Private Limited. ISBN: 978-981-13-1782-8. https://doi.org/10.1007/978-981-13-1783-5.

Li, L. (2018b). Modeling the monotonic and cyclic tensile stress-strain behavior of 2D and 2.5D woven C/SiC ceramic-matrix composites. *Mechanics of Composite Materials* 54: 165–178. https://doi.org/10.1007/s11029-018-9729-5.

Li, L. (2019a). A thermomechanical fatigue hysteresis-based damage evolution model for fiber-reinforced ceramic-matrix composites. *International Journal of Damage Mechanics* 28: 380–403. https://doi.org/10.1177/1056789518772162.

Li, L. (2019b). Micromechanics modeling of fatigue hysteresis behavior in carbon fiber-reinforced ceramic-matrix composites. Part I: Theoretical analysis. *Composites Part B Engineering* 159: 502–513. https://doi.org/10.1016/j.compositesb.2014.09.036.

Li, L. (2019c). Damage and fracture of fiber-reinforced ceramic-matrix composites under thermal fatigue loading in oxidizing atmosphere. *Journal of the Ceramic Society of Japan* 127: 67–80. http://doi.org/10.2109/jcersj2.18107.

Li, L. (2020a). Effect of interface properties on tensile and fatigue behavior of 2D woven SiC/SiC fiber-reinforced ceramic-matrix composites. *Advances in Materials Science and Engineering*: 3618984, 17 pp. https://doi.org/10.1155/2020/3618984.

Li, L. (2020b). Modeling tensile damage and fracture processes of fiber-reinforced ceramic-matrix composites under the effect of pre-exposure at elevated temperatures. *Ceramics - Silikaty* 64: 50–62. https://doi.org/10.13168/cs.2019.0048.

Li, L., Song, Y., and Sun, Y. (2013). Modeling the tensile behavior of unidirectional C/SiC ceramic-matrix composites. *Mechanics of Composite Materials* 49: 659–672. https://doi.org/10.1007/s11029-013-9382-y.

Li, L., Song, Y., and Sun, Y. (2015). Modeling the tensile behavior of cross-ply C/SiC ceramic-matrix composites. *Mechanics of Composite Materials* 51: 359–376. https://doi.org/10.1007/s11029-015-9507-6.

Lino Alves, F.J., Baptista, A.M., and Marques, A.T. (2016). Metal and ceramic matrix composites in aerospace engineering. In: *Advanced Composite Materials for Aerospace Engineering Processing, Properties and Applications* (eds. S. Rana and R. Fangueiro). Woodhead Publishing. ISBN: 978-0-08-100939-0. https://doi.org/10.1016/B978-0-08-100037-3.00003-1.

Mall, S. and Engesser, J.M. (2006). Effects of frequency on fatigue behavior of CVI C/SiC at elevated temperature. *Composites Science and Technology* 66: 863–874. https://doi.org/10.1016/j.compscitech.2005.06.020.

Morscher, G.N. and Baker, C. (2014). Electrical resistance of SiC fiber reinforced SiC/Si matrix composites at room temperature during tensile testing. *International Journal of Applied Ceramic Technology* 11: 263–272. https://doi.org/10.1111/ijac.12175.

Morscher, G.N., Sing, M., Kiser, J.D. et al. (2007). Modeling stress-dependent matrix cracking and stress-strain behavior in 2D woven SiC fiber reinforced CVI SiC composites. *Composites Science and Technology* 67: 1009–1017. https://doi.org/10.1016/j.compscitech.2006.06.007.

Naslain, R. (2004). Design, preparation and properties of non-oxide CMCs for application in engines and nuclear reactors: an overview. *Composites Science and Technology* 64: 155–170. https://doi.org/10.1016/S0266-3538(03)00230-6.

Naslain, R. (2005). SiC-matrix composites: nonbrittle ceramics for thermo-structural application. *International Journal of Applied Ceramic Technology* 2: 75–84. https://doi.org/10.1111/j.1744-7402.2005.02009.x.

Newton, C.D., Jones, J.P., Gale, L., and Bache, M.R. (2018). Detection of strain and damage distribution in SiC/SiC mechanical test coupons. *ASME Turbo Expo 2018: Turbomachinery Technical Conference and Exposition*, Oslo, Norway (11–15 June 2018). https://doi.org/10.1115/GT2018-75791.

Opalski, F.A. and Mall, S. (1994). Tension-compression fatigue behavior of a silicon carbide calcium-aluminosilicate ceramic matrix composites. *Journal of Reinforced Plastics and Composites* 13: 420–438. https://doi.org/10.1177/073168449401300503.

Rebillat, F. (2014). Advances in self-healing ceramic matrix composites. In: *Advances in Ceramic Matrix Composites* (ed. I.M. Low). Woodhead Publishing. ISBN: 978-0-85709-120-8. https://doi.org/10.1533/9780857098825.2.369.

Reynaud, P. (1996). Cyclic fatigue of ceramic-matrix composites at ambient and elevated temperatures. *Composites Science and Technology* 56: 809–814. https://doi.org/10.1016/0266-3538(96)00025-5.

Rouby, D. and Reynaud, P. (1993). Fatigue behavior related to interface modification during load cycling in ceramic-matrix fiber composites. *Composites Science and Technology* 48: 109–118. https://doi.org/10.1016/0266-3538(93)90126-2.

Ruggles-Wrenn, M.B., Christensen, D.T., Chamberlain, A.L. et al. (2011). Effect of frequency and environment on fatigue behavior of a CVI SiC/SiC ceramic matrix composite at 1200 °C. *Composites Science and Technology* 71: 190–196. https://doi.org/10.1016/j.compscitech.2010.11.008.

Sauder, C., Brusson, A., and Lamon, J. (2010). Influence of interface characteristics on the mechanical properties of Hi-Nicalon type-S or Tyranno-SA3 fiber-reinforced SiC/SiC minicomposites. *International Journal of Applied Ceramic Society* 7: 291–303. https://doi.org/10.1111/j.1744-7402.2010.02485.x.

Shuler, S.F., Holmes, J.W., and Wu, X. (1993). Influence of loading frequency on the room-temperature fatigue of a carbon-fiber/SiC-matrix composite. *Journal of the American Ceramic Society* 76: 2327–2336. https://doi.org/10.1111/j.1151-2916.1993.tb07772.x.

Staehler, J.M., Mall, S., and Zawada, L.P. (2003). Frequency dependence of high-cycle fatigue behavior of CVI C/SiC at room temperature. *Composites Science and Technology* 63: 2121–2131. https://doi.org/10.1016/S0266-3538(03)00190-8.

Vagaggini, E., Domergue, J.C., and Evans, A.G. (1995). Relationships between hysteresis measurements and the constituent properties of ceramic matrix composites: I. Theory. *Journal of the American Ceramic Society* 78: 2709–2720. https://doi.org/10.1111/j.1151-2916.1995.tb08046.x.

Wang, Y., Zhang, L., and Cheng, L. (2013). Comparison of tensile behaviors of carbon/ceramic composites with various fiber architectures. *International Journal of Applied Ceramic Technology* 10: 266–275. https://doi.org/10.1111/j.1744-7402.2011.02727.x.

Watanabe, F. and Manabe, T. (2018). Engine testing for the demonstration of a 3D-woven based ceramic matrix composite turbine vane design concept. *ASME Turbo Expo 2018: Turbomachinery Technical Conference and Exposition*, Oslo, Norway (11–15 June 2018). https://doi.org/10.1115/GT2018-75446.

Xia, Z. and Curtin, W.A. (2000). Tough-to-brittle transitions in ceramic-matrix composites with increasing interfacial shear stress. *Acta Materialia* 48: 4879–4892. https://doi.org/10.1016/S1359-6454(00)00291-3.

Xia, Z. and Li, L. (2014). Understanding interfaces and mechanical properties of ceramic matrix composites. In: *Advances in Ceramic Matrix Composites* (ed. I.M. Low). Woodhead Publishing. ISBN: 978-0-85709-120-8. https://doi.org/10.1533/9780857098825.2.267.

Yang, F.S. (2011). Research on fatigue behavior of 2.5D woven ceramic matrix composites. Master thesis. Nanjing University of Aeronautics and Astronautics, Nanjing.

Zawada, L.P., Butkus, L.M., and Hartman, G.A. (1991). Tensile and fatigue behavior of silicon carbide fiber-reinforced aluminosilicate glass. *Journal of the American Ceramic Society* 74: 2851–2858. https://doi.org/10.1111/j.1151-2916.1991.tb06854.x.

Zhang, C., Wang, X.W., Liu, Y.S. et al. (2013). Tensile fatigue of a 2.5D-C/SiC composite at room temperature and 900 °C. *Materials and Design* 49: 814–819. https://doi.org/10.1016/j.matdes.2013.01.076.

Zhang, C., Zhao, M., Liu, Y. et al. (2016). Tensile strength degradation of a 2.5D-C/SiC composite under thermal cycles in air. *Journal of the European Ceramic Society* 36: 3011–3019. https://doi.org/10.1016/j.jeurceramsoc.2015.12.007.

Zhu, S., Mizuno, M., Kagawa, Y., and Mutoh, Y. (1999). Monotonic tension, fatigue and creep behavior of SiC-fiber-reinforced SiC-matrix composites: a review. *Composites Science and Technology* 59: 833–851. https://doi.org/10.1016/S0266-3538(99)00014-7.

Zok, F.W. (2016). Ceramic-matrix composites enable revolutionary gains in turbine engine efficiency. *American Ceramic Society Bulletin* 95: 22–28.

3

Interface Assessment of Ceramic-Matrix Composites

3.1 Introduction

Ceramic materials possess high strength and modulus at elevated temperatures. But their use as structural components is severely limited because of their brittleness. The continuous fiber-reinforced ceramic-matrix composites (CMCs), by incorporating fibers in ceramic matrices, however, not only exploit their attractive high-temperature strength but also reduce the propensity for catastrophic failure. The carbon fiber-reinforced silicon carbide ceramic-matrix composites (C/SiC CMCs) are promising candidate high-temperature materials for application in a wide range of aerospace applications, such as hot section components of gas turbines (Naslain 2004), aero engines (Naslain 2005), shuttle nose (Krenkel and Berndt 2005), and thermal protection system (TPS) (Valentine et al. 2003).

Under fatigue loading, the damage in forms of matrix cracking, fiber/matrix interface debonding, interface wear, and interface oxidation may occur in CMCs during services (Reynaud et al. 1998; Staehler et al. 2003; Ruggles-Wrenn et al. 2012). The stress–strain hysteresis loops developed due to frictional slip occurred along any interface debonded region under fatigue loading can reveal composite's internal damages (Rouby and Reynaud 1993; Evans et al. 1995; Dassios et al. 2013). Holmes and Cho (1992) firstly performed an investigation on the effects of fatigue loading history and microstructural damage on fatigue hysteresis loops of unidirectional CMCs at room temperature. Shuler et al. (1993) investigated the effect of loading frequencies ranging from 1 to 85 Hz on fatigue behavior of 2D C/SiC composite. With the increase of loading frequencies, the extent of frictional heating increases. Under long-duration fatigue loading, the surface temperature rises at first and then gradually decreases, the fatigue hysteresis loops area also decrease gradually, which supports the gradual wear of fiber/matrix interface. Reynaud (1996) investigated the fatigue hysteresis loops evolution of two different types of CMCs at elevated temperatures in inert atmosphere. First, the fatigue hysteresis loops area of 2D SiC/SiC composite increases with the number of cycles increasing due to the interface radial thermal residual compressive stress. In the second ceramic composite, $[0/90]_s$ SiC/MAS-L, the fatigue hysteresis loops area decreases with the number of cycles increasing due to the interface radial thermal residual tensile stress. Mall and Engesser (2006) investigated fatigue behavior of 2D C/SiC composite at an elevated temperature of 550 °C under loading frequencies ranging from 0.1 to 375 Hz. The oxidation of carbon fiber is the major difference between room and elevated temperatures, which caused fatigue life reduction at lower frequencies at elevated temperature compared with that at room temperature (Staehler et al. 2003). However, at the loading frequency of 375 Hz, the oxidation of carbon fiber was almost absent or negligible at elevated temperature, due to the temperature rising during frictional heating caused by interface slip between fiber and matrix. Mei and Cheng (2009) investigated the cyclic loading/unloading hysteresis behavior of needled, 2D, 2.5D, and 3D C/SiC composites at room temperature. The increase in permanent strain and decrease in stiffness are affected by the fiber volume fraction in the loading direction.

The stress–strain hysteresis loop is an indicative of energy-dissipating mechanism and evolves from the interface slip mechanism. The intermittent unloading/reloading hysteresis loops under a mechanical test can be used as a useful tool to reveal instantaneous material response and damage evolution. Many researchers investigated the cyclic loading/unloading or fatigue hysteresis behavior of CMCs. Kotil et al. (1990) firstly performed an investigation on the effect of the interface shear stress on shape and area of cyclic loading/unloading hysteresis loops in unidirectional CMCs. Pryce and Smith (1993) investigated the cyclic loading/unloading hysteresis loops in unidirectional CMCs when the interface partially debonded based on the assumption of purely frictional loads transfer between fiber and matrix. Solti et al. (1995) investigated the cyclic loading/unloading hysteresis loops in unidirectional and cross-ply CMCs when the interface was chemically bonded and partially debonded by adopting the maximum interface shear strength criterion to determine the interface debonded length. Vagaggini et al. (1995) developed the cyclic loading/unloading hysteresis loops models of unidirectional CMCs based on the Hutchinson–Jensen fiber pull-out model (Hutchinson and Jensen 1990). Domergue et al. (1996) developed a methodology for evaluating the constituent properties of cross-ply CMCs from hysteresis loops measurements based on the Vagaggini's unidirectional hysteresis loops models (Vagaggini et al. 1995). Ahn and Curtin (1997) investigated the effect of matrix stochastic cracking on cyclic loading/unloading hysteresis loops in unidirectional CMCs by assuming the two-parameter Weibull distribution of matrix flaw and compared with Pryce–Smith model (Pryce and Smith 1993). Li (2013a) investigated the effects of the interface debonding, fiber Poisson contraction, fiber failure, and multiple matrices cracking on cyclic loading/unloading or fatigue hysteresis loops in unidirectional or cross-ply CMCs when the interface was chemically bonded. Fantozzi and Reynaud (2009) investigated the fatigue hysteresis behavior of bi- or multi-directional (cross-weave, cross-ply, 2.5D, $[0/+60/-60]_n$) with SiC or C long fibers reinforced SiC, MAS-L, Si–B–C, or C matrix at room and elevated temperatures in inert and oxidation conditions. By assuming the mechanical hysteresis behavior of composite is mainly controlled by the interface frictional slip in longitudinal yarns, the fatigue hysteresis loops shape evolution of these composites has been analyzed.

The fiber/matrix interface shear stress is a key parameter in fatigue hysteresis behavior of CMCs, which affects the fatigue energy lost, i.e. the area of fatigue hysteresis loops. Under fatigue loading at room temperature, the slip displacements between fiber and matrix could reduce the interface shear stress (Rouby and Reynaud 1993; Evans et al. 1995). Evidences of the interface wear that a reduction in the height of asperities occurs along the fiber coating for different thermal misfit, surface roughness, and frictional sliding velocity have been presented by push-out and push-back tests on a ceramic composite system (Rouby and Louet 2002). The interface wear process can be facilitated by temperature rising that occurs along fiber/matrix interface, as frictional dissipation proceeds (Holmes and Cho 1992; Kim and Liaw 2005; Liu et al. 2008), i.e. the temperature rising exceeded 100 K under fatigue loading at 75 Hz between stress levels of 220 and 10 MPa in unidirectional SiC/CAS-II composite (Holmes and Cho 1992). Under fatigue loading at elevated temperatures in air atmosphere, the interphase would react to form CO if the fiber coating is carbon or PyC, resulting in a large reduction in the interface shear stress (Holmes and Sørensen 1995). Evidences of the interface oxidation, i.e. a uniformly reduction in fiber diameter and a longer fiber pullout length occurs in a 2D C/SiC composite, have been presented by a non-stress oxidation experiment at an elevated temperature of 700 °C in air atmosphere (Yang et al. 2009), and a tensile fatigue experiment at an elevated temperature of 550 °C in air atmosphere (Mall and Engesser 2006). Moevus et al. (2006) investigated the static fatigue behavior of 2.5D C/[Si–B–C] composite at an elevated temperature of 1200 °C in air atmosphere. The hysteresis loops area after a static fatigue of 144 hours under a steady stress of 170 MPa, significantly decreased, attributed to a decrease of the interface shear stress caused by PyC interphase recession by oxidation.

The fiber/matrix interface shear stress of CMCs under fatigue loading can be estimated from hysteresis loops (Evans et al. 1995; Reynaud 1996; Domergue et al. 1995, 1996; Fantozzi and Reynaud 2009). Cho et al. (1991) developed an approach to estimate the interface shear stress from frictional heating measurement. By analyzing the frictional heating data, Holmes and Cho (1992) found that the interfacial shear stress of unidirectional SiC/CAS-II composite undergoes an initially rapid decrease at the initial stages of fatigue loading, i.e. from an initial value of over 20 MPa, to approximately 5 MPa after 25 000 cycles. Evans et al. (1995) developed an approach to evaluate the interface shear stress by analyzing parabolic regions of hysteresis loops based on the Vagaggini's hysteresis loops models (Vagaggini et al. 1995). The initial interface shear stress of unidirectional SiC/CAS composite was approximately 20 MPa, and degraded to about 5 MPa at the 30th cycle. Solti et al. (2000) proposed a means of inferring the state of the interface through comparison of experimental and theoretical fatigue hysteresis loss energy on a cycle by cycle base. Li and Song (2010, 2013) and Li (2013b, 2014, 2016) established the theoretical relationship between the fatigue hysteresis loss energy and the interface shear stress of unidirectional, cross-ply, and 2.5D CMCs.

In this chapter, the relationship between the hysteresis dissipated energy and temperature rising of the external surface in fiber-reinforced CMCs under cyclic loading is analyzed. Based on the fatigue hysteresis theories considering fiber failure, the hysteresis dissipated energy and a hysteresis dissipated energy-based damage parameter changing with the increase of cycle number are investigated.

The experimental temperature rise-based damage parameter of unidirectional SiC/CAS-II, cross-ply SiC/CAS, and 2D C/SiC composites corresponding to different fatigue peak stresses and cycle numbers are predicted. The fatigue hysteresis behavior of unidirectional, cross-ply, and 2.5D C/SiC composites at room temperature and 800 °C in air atmosphere are investigated. Comparing experimental fatigue hysteresis dissipated energy with theoretical computational values, the evolution of the interface shear stress with the number of cycles increasing is estimated.

3.2 Relationships Between Interface Slip and Temperature Rising in CMCs

In this section, the relationship between hysteresis dissipated energy and temperature rising of the external surface in fiber-reinforced CMCs during the application of cyclic loading has been analyzed. The temperature rise, which is caused by frictional slip of fibers within the composite, is related to the hysteresis dissipated energy. Based on the fatigue hysteresis theories considering fibers failure, the hysteresis dissipated energy and a hysteresis dissipated energy-based damage parameter changing with the increase of cycle number have been investigated. The relationship between the hysteresis dissipated energy, a hysteresis dissipated energy-based damage parameter, and a temperature rise-based damage parameter has been established. The experimental temperature rise-based damage parameter of unidirectional, cross-ply, and 2D woven CMCs corresponding to different fatigue peak stresses and cycle numbers has been predicted. It was found that the temperature rise-based parameter can be used to monitor the fatigue damage evolution and predict the fatigue life of fiber-reinforced CMCs.

3.2.1 Hysteresis Theories

If matrix multicracking and fiber/matrix interface debonding are present upon first loading, the stress–strain hysteresis loops develop as a result of energy dissipation through frictional slip between fibers and matrix upon unloading and subsequent reloading. Upon unloading, the counter slip occurs in the interface debonded region. The interface debonded region can be divided into two regions, i.e. the interface counter slip region and interface slip region, as shown in Figure 3.1a. The interface counter slip length is denoted to be y. Upon reloading, the new slip occurs in the interface debonded region. The interface debonded region can be divided into three regions, i.e. the interface new slip region, interface counter slip region, and interface slip region, as shown in Figure 3.1b. The interface new slip region is denoted to be z. When fibers failure occurs under fatigue loading, the slip of intact and broken fibers in the interface debonded region would affect the hysteresis dissipated energy.

Based on the damage mechanism of interface frictional slip between fibers and matrix upon unloading/reloading, the stress–strain hysteresis loops considering

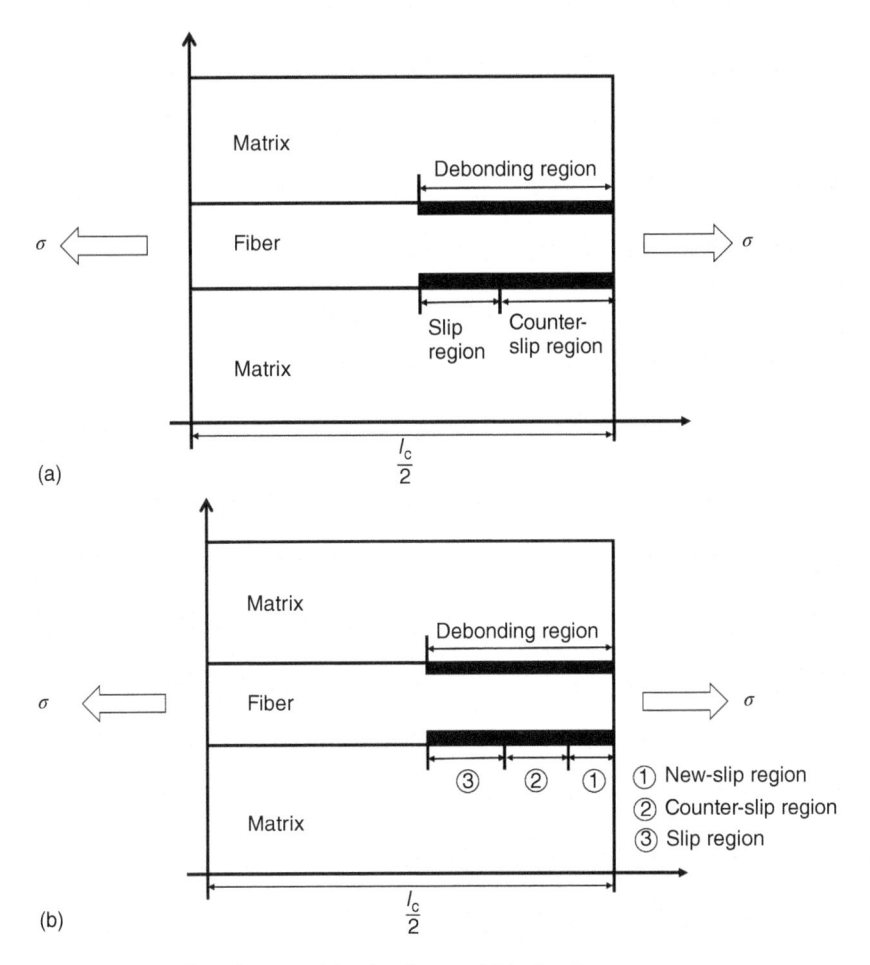

Figure 3.1 Interface slip upon (a) unloading; and (b) reloading.

fiber failure can be classified into four different cases, as follows:

(1) Case I. The interface partial debonds and the fiber complete slides relative to the matrix.
(2) Case II. The interface partial debonds and the fiber partial slides relative to the matrix;
(3) Case III. The interface complete debonds and the fiber partial slides relative to the matrix.
(4) Case IV. The interface complete debonds and the fiber complete slides relative to the matrix.

3.2.1.1 Case I

Upon unloading to σ_{tr_pu} ($\sigma_{tr\_pu} > \sigma_{min}$), the interface counter slip length y approaches the interface debonding length l_d, i.e. $y(\sigma = \sigma_{tr\_pu}) = l_d$. Upon continually unloading, the interface counter slip length would not change, i.e. $y(\sigma_{min}) = y(\sigma_{tr\_pu})$. The fiber axial stress distribution upon unloading is shown

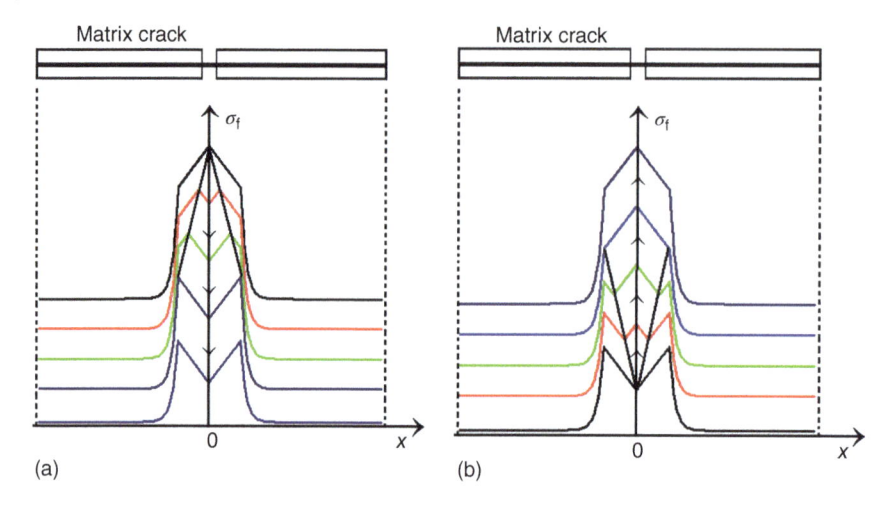

3 Interface Assessment of Ceramic-Matrix Composites

Figure 3.2 The fiber axial stress distribution of Case I upon (a) unloading; and (b) reloading.

in Figure 3.2a. The interface counter slip length y is determined by fracture mechanics approach.

$$y = \frac{1}{2}\left\{ l_d - \left[\frac{r_f}{2}\left(\frac{V_m E_m}{E_c \tau_i}\Phi_U - \frac{1}{\rho}\right) \right.\right.$$

$$\left.\left. - \sqrt{\left(\frac{r_f}{2\rho}\right)^2 - \frac{r_f^2 V_f V_m E_f E_m \Phi_U}{4 E_c^2 \tau_i^2}\left(\Phi_U - \frac{\sigma}{V_f}\right) + \frac{r_f V_m E_m E_f}{E_c \tau_i^2}\zeta_d}\right]\right\} \quad (3.1)$$

where ρ denotes the shear-lag model parameter; and Φ_U denotes the stress carried by intact fiber at the matrix crack plane upon unloading, which is determined by the following relation:

$$\frac{\sigma}{V_f} = 2\Phi\left(\frac{\sigma_k}{\Phi}\right)^{m_f+1}\left\{\exp\left[-\left(\frac{\Phi-\Phi_U}{2\Phi}\right)\left(\frac{\Phi}{\sigma_k}\right)^{m_f+1}\right] - 1 + \frac{1}{2}q\right\} \quad (3.2)$$

The unloading stress–strain relationship is given by Eq. (3.3) and is divided into two regions, i.e. (i) when $\sigma > \sigma_{tr\_pu}$, the unloading strain is determined by Eq. (3.3); and (ii) when $\sigma < \sigma_{tr\_pu}$, the unloading strain is determined by Eq. (3.3) by setting $y = l_d$.

$$\varepsilon_{unloading} = \frac{\Phi_U + \Delta\sigma}{E_f} + 4\frac{\tau_i}{E_f}\frac{y^2}{r_f l_c} - \frac{\tau_i}{E_f}\frac{(2y-l_d)(2y+l_d-l_c)}{r_f l_c} - (\alpha_c - \alpha_f)\Delta T \quad (3.3)$$

where α_f and α_c denote the fiber and composite thermal expansion coefficient, respectively; $\Delta\sigma$ denotes the additional stress in intact fibers resulting from gross slip of adjacent fractured fibers; ΔT denotes the temperature difference between the fabricated temperature T_0 and room temperature T_1 ($\Delta T = T_1 - T_0$).

Upon reloading to σ_{tr_pr} ($\sigma_{tr\_pr} < \sigma_{max}$), the interface new slip length z approaches interface debonding length l_d, i.e. $z(\sigma = \sigma_{tr\_pr}) = l_d$. Upon continually reloading, the interface new slip length would not change, i.e. $z(\sigma_{max}) = z(\sigma_{tr\_pr})$. The fiber axial stress distribution upon reloading is shown in Figure 3.2b. The reloading interface new slip length is determined by fracture mechanics approach.

$$z = y(\sigma_{min}) - \frac{1}{2}\left\{ l_d - \left[\frac{r_f}{2}\left(\frac{V_m E_m}{E_c \tau_i} \Phi_R - \frac{1}{\rho} \right) \right.\right.$$

$$\left.\left. - \sqrt{\left(\frac{r_f}{2\rho} \right)^2 - \frac{r_f^2 V_f V_m E_f E_m \Phi_R}{4 E_c^2 \tau_i^2}\left(\Phi_R - \frac{\sigma}{V_f} \right) + \frac{r_f V_m E_m E_f}{E_c \tau_i^2}\zeta_d} \right]\right\} \quad (3.4)$$

The reloading stress–strain relationship is determined by Eq. (3.5) and is divided into two regions, i.e. (i) when $\sigma < \sigma_{tr\_pr}$, the reloading strain is determined by Eq. (3.5); and (ii) when $\sigma > \sigma_{tr\_pr}$, the reloading strain is determined by Eq. (3.5) by setting $z = l_d$.

$$\varepsilon_{reloading} = \frac{\Phi_R + \Delta\sigma}{E_f} - 4\frac{\tau_i}{E_f}\frac{z^2}{r_f l_c} + 4\frac{\tau_i}{E_f}\frac{(y-2z)^2}{r_f l_c}$$
$$+ 2\frac{\tau_i}{E_f}\frac{(l_d - 2y + 2z)(l_d + 2y - 2z - l_c)}{r_f l_c} - (\alpha_c - \alpha_f)\Delta T \quad (3.5)$$

where Φ_R denotes the stress carried by intact fiber at the matrix crack plane upon reloading, which is determined by the following relation:

$$\frac{\sigma}{V_f} = 2\Phi\left(\frac{\sigma_k}{\Phi} \right)^{m_f+1}\left\{ \exp\left[-\left(\frac{\Phi_m}{2\Phi} \right)\left(\frac{\Phi}{\sigma_k} \right)^{m_f+1} \right]\right.$$
$$\left. - \exp\left[-\left(\frac{\Phi_R - \Phi + \Phi_m}{2T} \right)\left(\frac{\Phi}{\sigma_k} \right)^{m_f+1} \right] + \frac{1}{2}q \right\} \quad (3.6)$$

In Eq. (3.6), Φ_m is determined by the following equation:

$$0 = 2\Phi\left(\frac{\sigma_k}{\Phi} \right)^{m_f+1}\left\{ \exp\left[-\left(\frac{\Phi_m}{2\Phi} \right)\left(\frac{\Phi}{\sigma_k} \right)^{m_f+1} \right] - 1 + \frac{1}{2}q \right\} \quad (3.7)$$

3.2.1.2 Case II

Upon complete unloading, the interface counter slip length y is less than the interface debonding length l_d, i.e. $y(\sigma_{min}) < l_d$. The unloading interface counter slip length y is given by Eq. (3.1). The fiber axial stress distribution upon unloading is shown in Figure 3.3a. The unloading stress–strain relationship is determined by Eq. (3.3). Upon reloading, the interface new slip length z is less than the interface debonded length l_d, i.e. $z(\sigma_{max}) < l_d$. The reloading interface new slip length z is given by Eq. (3.4). The fiber axial stress distribution upon reloading is illustrated in Figure 3.3b. The reloading stress–strain relationship is determined by Eq. (3.5).

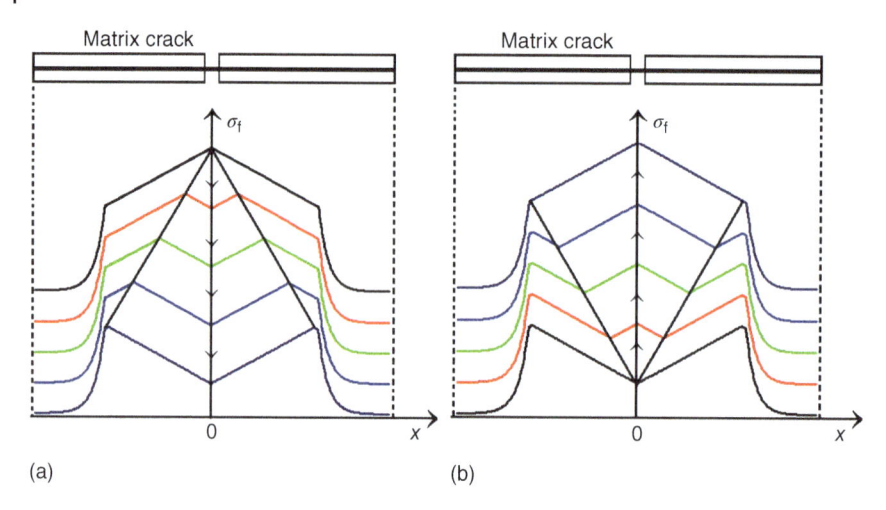

Figure 3.3 The fiber axial stress distribution of Case II upon (a) unloading; and (b) reloading.

3.2.1.3 Case III

When the interface complete debonds, the interface counter slip length is given by Eq. (3.8). Upon complete unloading, the interface counter slip length y is less than half matrix crack spacing of $l_c/2$, i.e. $y(\sigma_{\min}) < l_c/2$.

$$y = \frac{r_f}{4\tau_i}\left[(\Phi - \Phi_U) - \frac{E_f}{E_c}(\sigma_{\max} - \sigma)\right] \tag{3.8}$$

The fiber axial stress distribution upon unloading is shown in Figure 3.4a. The unloading stress–strain relationship is determined by the following equation.

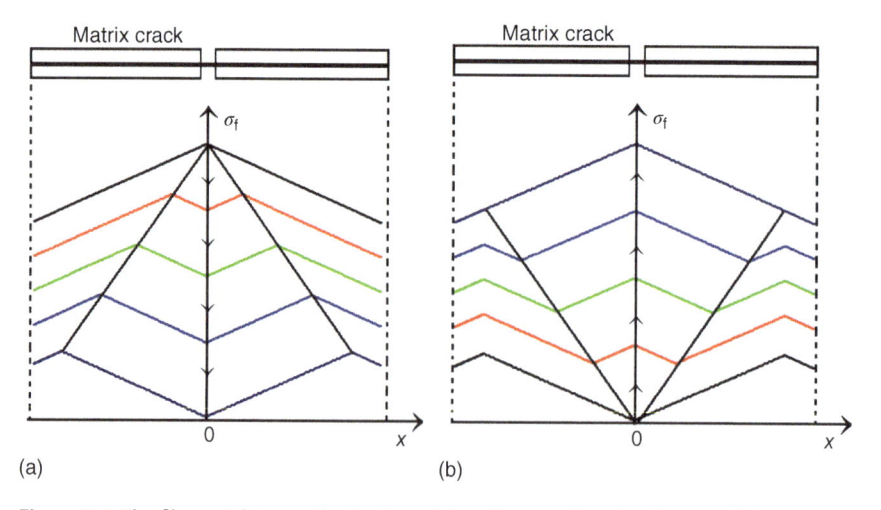

$$\varepsilon_{\text{unloading}} = \frac{\Phi_U + \Delta\sigma}{E_f} + 4\frac{\tau_i}{E_f}\frac{y^2}{r_f l_c} - 2\frac{\tau_i}{E_f}\frac{(2y - l_c/2)^2}{r_f l_c} - (\alpha_c - \alpha_f)\Delta T \tag{3.9}$$

Figure 3.4 The fiber axial stress distribution of Case III upon (a) unloading; and (b) reloading.

Upon reloading, the interface new slip length z is given by Eq. (3.10). Upon reloading to σ_{max}, the interface new slip length z is less than half matrix crack spacing of $l_c/2$, i.e. $z(\sigma_{max}) < l_c/2$.

$$z = y(\sigma_{min}) - \frac{r_f}{4\tau_i}\left[(\Phi - \Phi_R) - \frac{E_f}{E_c}(\sigma_{max} - \sigma)\right] \quad (3.10)$$

The fiber axial stress distribution upon reloading is shown in Figure 3.4b. The reloading stress–strain relationship is determined by the following equation.

$$\varepsilon_{reloading} = \frac{\Phi_R + \Delta\sigma}{E_f} - 4\frac{\tau_i}{E_f}\frac{z^2}{r_f l_c} + 4\frac{\tau_i}{E_f}\frac{(y-2z)^2}{r_f l_c} - 2\frac{\tau_i}{E_f}\frac{(l_c/2 - 2y + 2z)^2}{r_f l_c}$$
$$- (\alpha_c - \alpha_f)\Delta T \quad (3.11)$$

3.2.1.4 Case IV

Upon unloading to σ_{tr_fu} ($\sigma_{tr\_fu} > \sigma_{min}$), the interface counter slip length y approaches half matrix crack spacing of $l_c/2$, i.e. $y(\sigma = \sigma_{tr\_fu}) = l_c/2$. When $\sigma > \sigma_{tr\_fu}$, the interface counter slip length y is less than half matrix crack spacing $l_c/2$, i.e. $y(\sigma > \sigma_{tr\_fu}) < l_c/2$. The interface counter slip length y is given by Eq. (3.8). When $\sigma_{min} < \sigma < \sigma_{tr\_fu}$, the interface counter slip occurs over the entire matrix crack spacing, i.e. $y(\sigma_{min} < \sigma < \sigma_{tr\_fu}) = l_c/2$. The fiber axial stress distribution upon unloading is shown in Figure 3.5a. The unloading stress–strain relationship is divided into two regions, i.e. (i) when $\sigma > \sigma_{tr\_fu}$, the unloading strain is determined by Eq. (3.9); and (ii) when $\sigma < \sigma_{tr\_fu}$, the unloading strain is determined by Eq. (3.9) by setting $y = l_c/2$.

Upon reloading to σ_{tr_fr} ($\sigma_{tr\_fr} < \sigma_{max}$), the interface new slip length z approaches half matrix crack spacing of $l_c/2$, i.e. $z(\sigma = \sigma_{tr\_fr}) = l_c/2$. When $\sigma < \sigma_{tr\_fr}$, the interface new slip length z is less than half matrix crack spacing of $l_c/2$, i.e. $z(\sigma < \sigma_{tr\_fr}) < l_c/2$. The interface new slip length z is given by Eq. (3.10). When $\sigma_{tr\_fr} < \sigma < \sigma_{max}$, the interface new slip length occupies the entire matrix

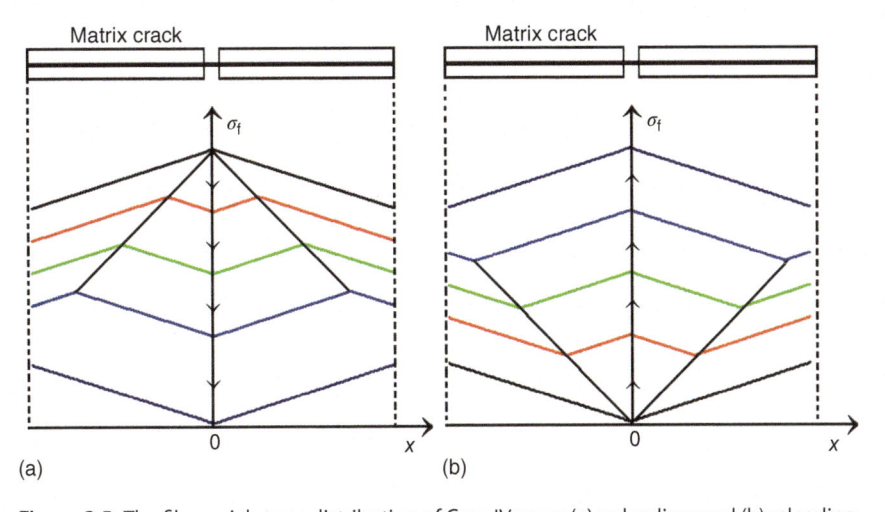

Figure 3.5 The fiber axial stress distribution of Case IV upon (a) unloading; and (b) reloading.

crack spacing of $l_c/2$, i.e. $z(\sigma_{tr\_fr} < \sigma < \sigma_{max}) = l_c/2$. The fiber axial stress distribution upon reloading is shown in Figure 3.5b. The reloading stress–strain relationship is divided into two regions, i.e. (i) when $\sigma < \sigma_{tr\_fr}$, the reloading strain is determined by Eq. (3.11); and (ii) when $\sigma > \sigma_{tr\_fr}$, the reloading strain is determined by Eq. (3.11) by setting $z = l_c/2$.

3.2.2 Experimental Comparisons

Under fatigue loading, the area associated with hysteresis loops is the dissipated energy during corresponding cycle, which is defined as,

$$U_e = \int_{\sigma_{min}}^{\sigma_{max}} [\varepsilon_{unloading}(\sigma) - \varepsilon_{reloading}(\sigma)] d\sigma \qquad (3.12)$$

where $\varepsilon_{unloading}$ denotes the unloading strain; and $\varepsilon_{reloading}$ denotes the reloading strain. Substituting unloading and reloading strains in Eqs. (3.3), (3.5), (3.9), and (3.11) into Eq. (3.12), the hysteresis dissipated energy corresponding to different interface slip cases can be obtained.

The hysteresis dissipated energy-based damage parameter Π is defined as

$$\Pi = \frac{U_e(N) - U_e(N=1)}{U_f(N_f) - U_e(N=1)} \qquad (3.13)$$

The temperature rise-based damage parameter Ψ is defined as

$$\Psi = \frac{T_n - T_{initial}}{T_f - T_{initial}} \qquad (3.14)$$

where T_n denotes the surface temperature at the nth cycle; $T_{initial}$ denotes the initial temperature rising at the first cycle; and T_f denotes the temperature rising at the failure cycle number.

3.2.2.1 Unidirectional CMCs

Holmes et al. (1994) investigated the frictional heating of unidirectional SiC/CAS-II composite under tension–tension fatigue loading at room temperature. The infrared pyrometers were used to measure the temperature increase of the specimen surface during fatigue experiments at sinusoidal frequency of $f = 350$ Hz. The fatigue peak stresses were $\sigma_{max} = 240$, 220, and 200 MPa and the valley stress was $\sigma_{min} = 10$ MPa. The maximum fatigue cycles were defined at $N = 5\,000\,000$ cycles. The basic materials properties of unidirectional SiC/CAS-II composite are given by the following: $V_f = 40\%$, $E_f = 190$ GPa, $E_m = 90$ GPa, $r_f = 7.5\,\mu m$, $\alpha_f = 4\times 10^{-6}\,°C$, $\alpha_m = 5\times 10^{-6}\,°C$, $\Delta T = -1000\,°C$, $\zeta_d = 0.4\,J/m^2$, $\tau_{io} = 25$ MPa, $\tau_{imin} = 5$ MPa, $\omega = 5\times 10^{-7}$, $\lambda = 1.3$, $p_1 = 0.01$, $p_2 = 1.0$, $\sigma_c = 2.0$ GPa, and $m_f = 2$.

When the fatigue peak stress is $\sigma_{max} = 240$ MPa, the stress–strain hysteresis loops corresponding to the cycle number of $N = 1$, 30 000, 50 000, 100 000, and 150 000 are shown in Figure 3.6a, corresponding to the interface slip Cases I, II, II, III, and IV, respectively. The fatigue hysteresis dissipated energy U_e versus applied cycle number curve is shown in Figure 3.6b, in which the fatigue hysteresis dissipated energy increases from $U_e = 15.4$ kPa at $N = 1$ to $U_e = 79.6$ kPa

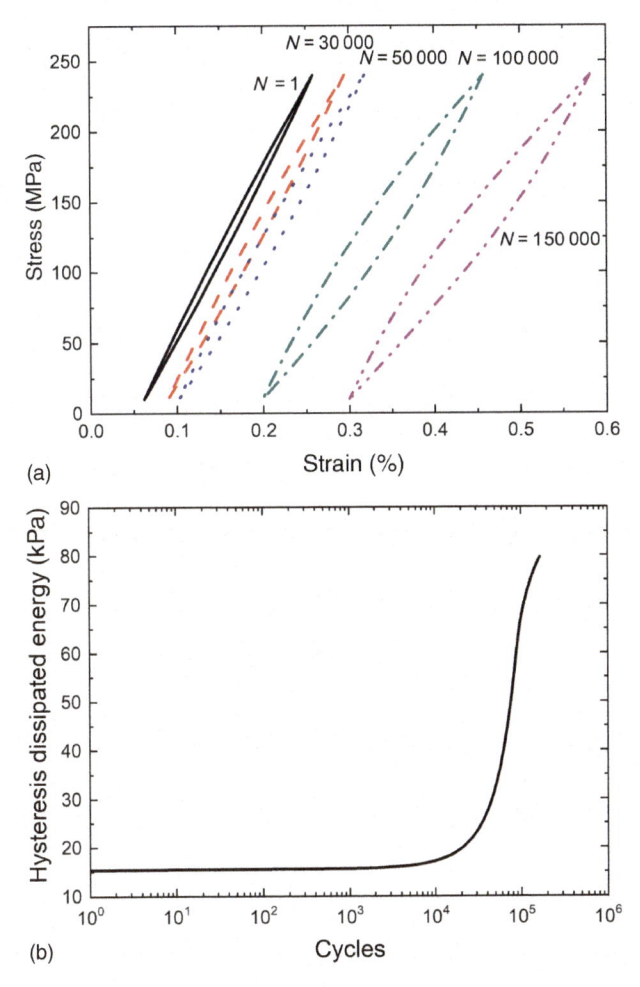

Figure 3.6 (a) The fatigue hysteresis loops corresponding to different applied cycle number; (b) the fatigue hysteresis dissipated energy versus applied cycle number; (c) the temperature increase ($T_n - T_{initial}$) versus applied cycle number; and (d) the hysteresis dissipated energy-based damage parameter Π and temperature rise-based damage parameter Ψ versus applied cycle number of unidirectional SiC/CAS-II composite under fatigue peak stress of $\sigma_{max} = 240$ MPa at room temperature.

at $N = 166\,460$. The temperature rising (i.e. $T_n - T_{initial}$) versus applied cycle number curve is shown in Figure 3.6c, in which the temperature rise increase from zero at $N = 1$ to 157.6 K at $N = 166\,460$. When the applied stress is sufficient to cause extensive damage of the composite, i.e. leading to fracture before run out, the internal damage would keep the temperature rising. The hysteresis dissipated energy-based damage parameter Π and the temperature rise-based damage parameter Ψ versus cycle number curves are given in Figure 3.6d, in which the evolution of hysteresis dissipated energy-based damage parameter Π is in coincidence with that of temperature rise-based damage parameter Ψ.

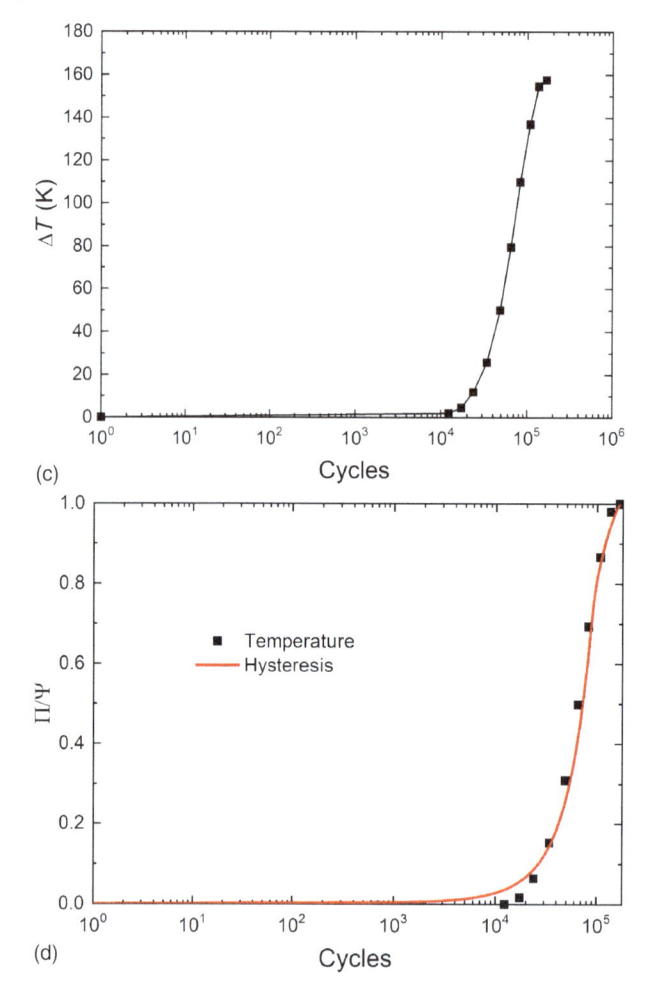

3 Interface Assessment of Ceramic-Matrix Composites

(c) [Graph: ΔT (K) vs Cycles, x-axis 10^0 to 10^6, y-axis 0 to 180]

(d) [Graph: Π/Ψ vs Cycles, showing Temperature (squares) and Hysteresis (line), x-axis 10^0 to 10^5, y-axis 0.0 to 1.0]

Figure 3.6 (Continued)

When the fatigue peak stress is $\sigma_{max} = 220$ MPa, the stress–strain hysteresis loops corresponding to the applied cycle number of $N = 1$, 50 000, 100 000, 150 000, and 200 000 are shown in Figure 3.7a, corresponding to the interface slip Cases I, II, II, III, and IV, respectively. The fatigue hysteresis dissipated energy U_e versus applied cycle number curve is shown in Figure 3.7b, in which the hysteresis dissipated energy increases from $U_e = 11.5$ kPa at $N = 1$ to $U_e = 59.8$ kPa at $N = 236\,640$. The temperature rising (i.e. $T_n - T_{initial}$) versus applied cycle number curve is shown in Figure 3.7c, in which the temperature rise increase from zero at $N = 1$ to 127.8 K at $N = 236\,640$. The fatigue hysteresis dissipated energy-based damage parameter Π and the temperature rise-based damage parameter Ψ versus applied cycle number curves are given in Figure 3.7d, in which the evolution of hysteresis dissipated energy-based damage parameter Π is in coincidence with that of temperature rise-based damage parameter Ψ.

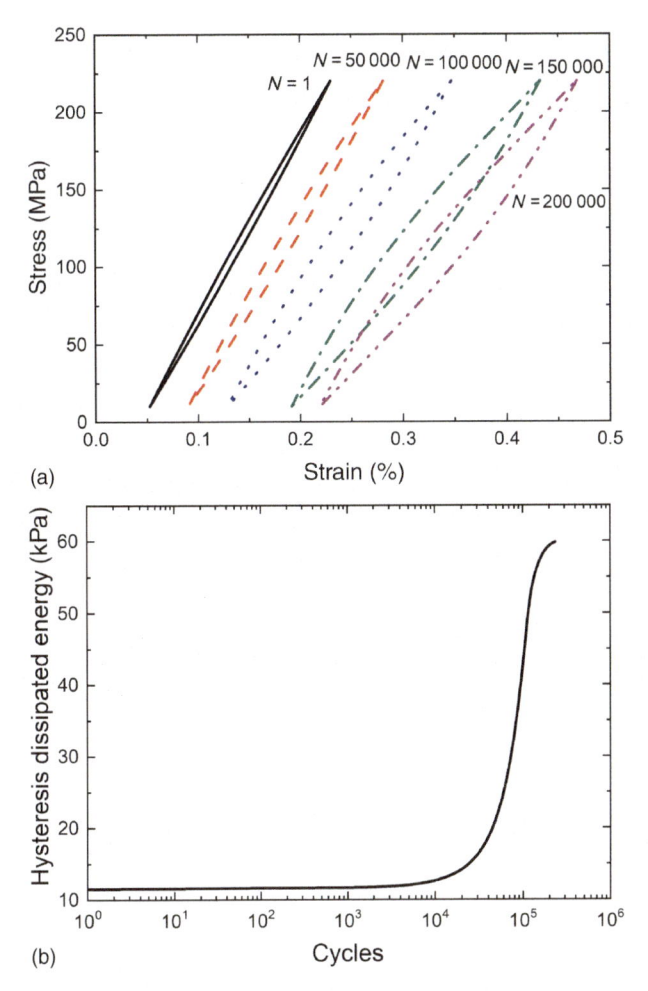

Figure 3.7 (a) The fatigue hysteresis loops corresponding to different applied cycle number; (b) the fatigue hysteresis dissipated energy versus applied cycle number; (c) the temperature increase ($T_n - T_{initial}$) versus applied cycle number; and (d) the hysteresis dissipated energy-based damage parameter Π and temperature rise-based damage parameter Ψ versus cycle number of unidirectional SiC/CAS-II composite under fatigue peak stress of $\sigma_{max} = 220$ MPa at room temperature.

When the fatigue peak stress is $\sigma_{max} = 200$ MPa, the stress–strain hysteresis loops corresponding to the applied cycle number of $N=1$, 100 000, 200 000, 300 000, 400 000, and 500 000 are shown in Figure 3.8a, corresponding to the interface slip Cases I, II, II, II, III, and IV, respectively. The fatigue hysteresis dissipated energy U_e versus applied cycle number curve is shown in Figure 3.8b, in which the fatigue hysteresis dissipated energy increases from $U_e = 8.5$ kPa at $N=1$ to $U_e = 51.2$ kPa at $N=511\,020$. The temperature increase (i.e. $T_n - T_{initial}$) versus applied cycle number curve is shown in Figure 3.8c, in which the temperature rise increase from zero at $N=1$ to 85.9 K at $N=511\,020$. The hysteresis dissipated energy-based damage parameter Π and temperature

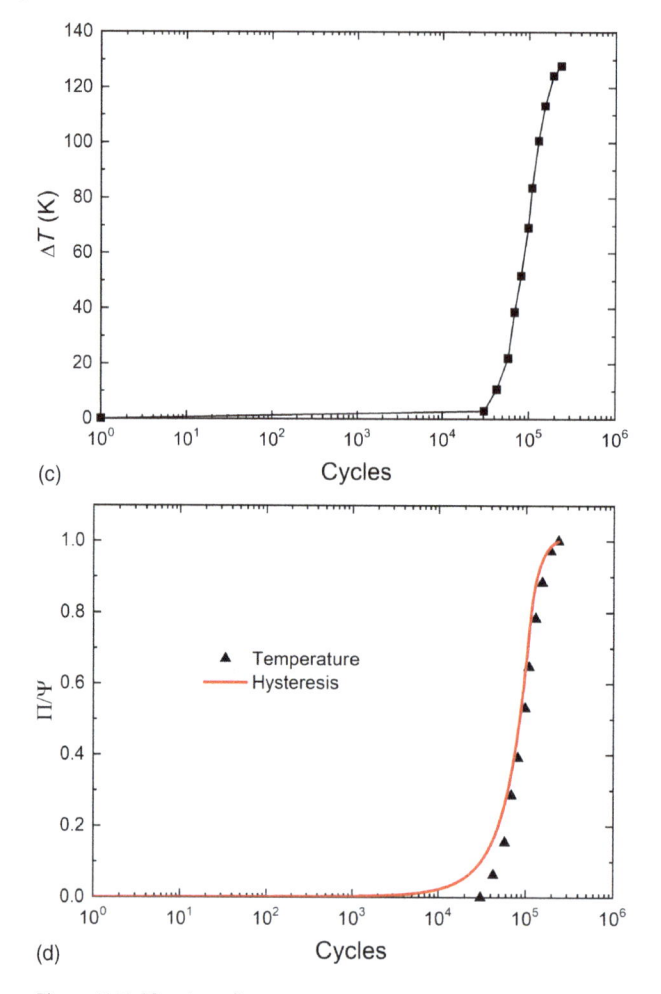

Figure 3.7 (Continued)

rise-based damage parameter Ψ versus applied cycle number curves are given in Figure 3.8d, in which the evolution of hysteresis dissipated energy-based damage parameter Π is in coincidence with that of temperature rise-based damage parameter Ψ.

3.2.2.2 Cross-Ply CMCs

Kim and Liaw (2005) investigated the tension–tension fatigue behavior of cross-ply SiC/CAS composite with the aid of infrared thermography at room temperature. The fatigue stress ratio was $R=0.1$ and the loading frequency was $f=20$ Hz. The fatigue peak stress was $\sigma_{max}=133$ MPa and the failure cycle number was $N=4283$. The basic materials properties of cross-ply SiC/CAS composite are given by the following: $V_f=40\%$, $E_f=192$ GPa, $E_m=98$ GPa $r_f=7.5$ μm, $\alpha_f=4\times10^{-6}$ °C, $\alpha_m=5\times10^{-6}$ °C, $\Delta T=-1000$ °C, $\zeta_d=0.4$ J/m^2,

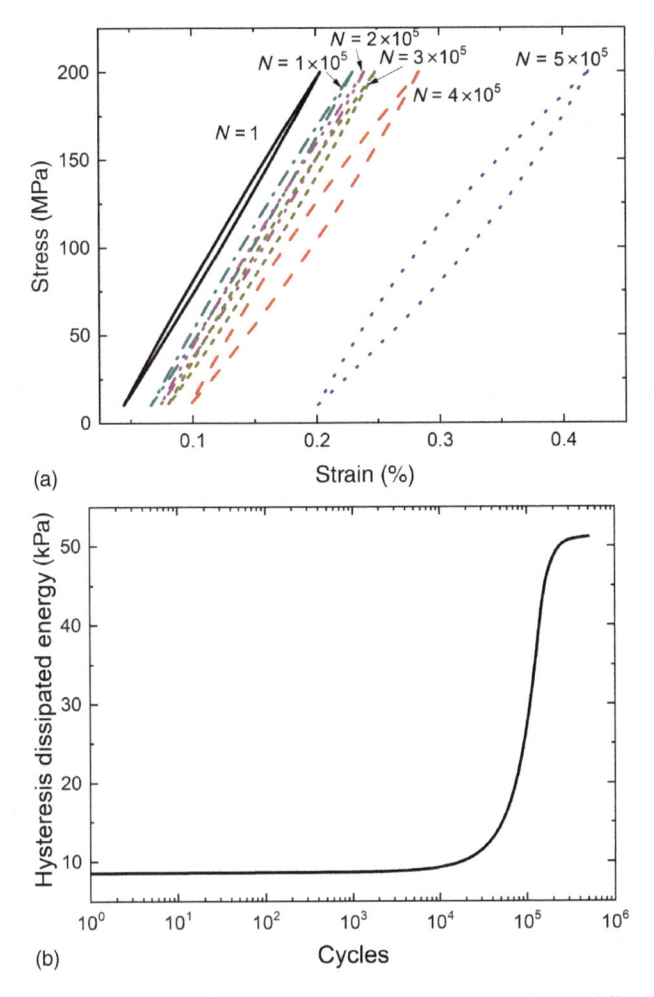

Figure 3.8 (a) The fatigue hysteresis loops corresponding to different applied cycle number; (b) the fatigue hysteresis dissipated energy versus applied cycle number; (c) the temperature increase ($T_n - T_{initial}$) versus applied cycle number; and (d) the hysteresis dissipated energy-based damage parameter Π and temperature rise-based damage parameter Ψ versus applied cycle number of unidirectional SiC/CAS-II composite under fatigue peak stress of $\sigma_{max} = 200$ MPa at room temperature.

$\tau_{io} = 25$ MPa, $\tau_{imin} = 5$ MPa, $\omega = 8 \times 10^{-5}$, $\lambda = 1.2$, $p_1 = 0.01$, $p_2 = 1.0$, $\sigma_c = 2.0$ GPa, and $m_f = 2$.

When the fatigue peak stress is $\sigma_{max} = 133$ MPa, the stress–strain hysteresis loops corresponding to the applied cycle number of $N = 1$, 1000, 2000, 3000, and 4000 are shown in Figure 3.9a, corresponding to the interface slip Cases I, II, III, IV, and IV, respectively. The fatigue hysteresis dissipated energy U_e versus applied cycle number curve is shown in Figure 3.9b, in which the fatigue hysteresis dissipated energy increases from $U_e = 20.9$ kPa at $N = 1$ to $U_e = 51.4$ kPa at $N = 4283$. The temperature increase (i.e. $T_n - T_{initial}$) versus applied cycle number curve is shown in Figure 3.9c, in which the temperature rise increase from zero at $N = 1$

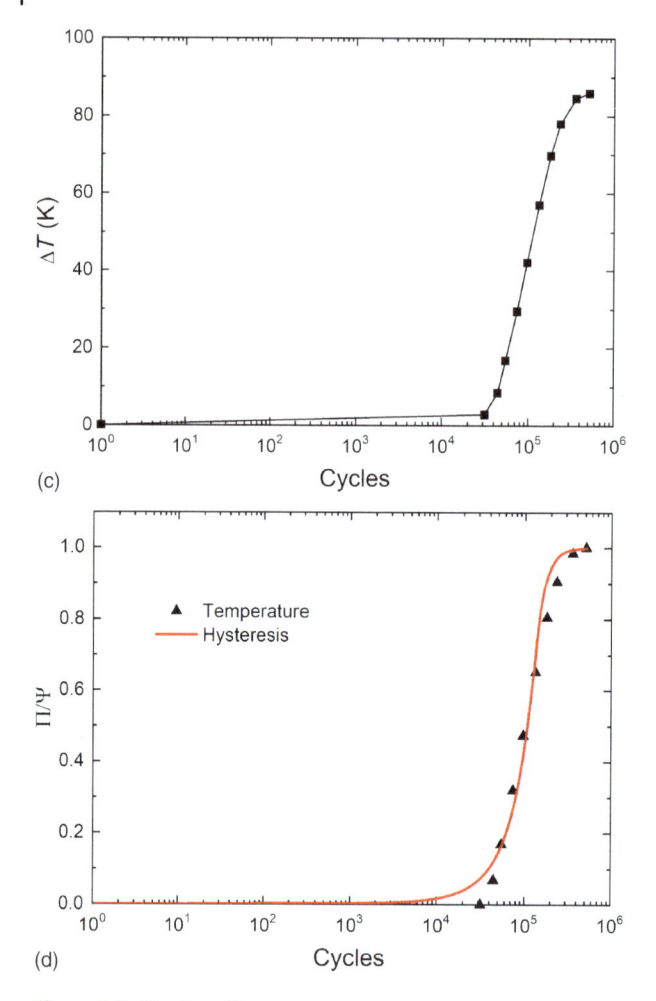

Figure 3.8 (Continued)

to 11.3 K at $N = 4200$. The hysteresis dissipated energy-based damage parameter Π and the temperature rise-based damage parameter Ψ versus cycle number curves are given in Figure 3.9d, in which the evolution of hysteresis dissipated energy-based damage parameter Π is in coincidence with that of temperature rise-based damage parameter Ψ.

3.2.2.3 2D CMCs

Shuler et al. (1993) investigated the tension–tension fatigue behavior of a 2D C/SiC composite at room temperature. The fatigue testing was performed under load control at a sinusoidal loading frequency of $f = 10\,\text{Hz}$ with a stress ratio of $R = 0.1$. The fatigue peak stress was $\sigma_{max} = 335\,\text{MPa}$ The basic materials properties of 2D C/SiC composite are given by the following: $V_f = 45\%$,

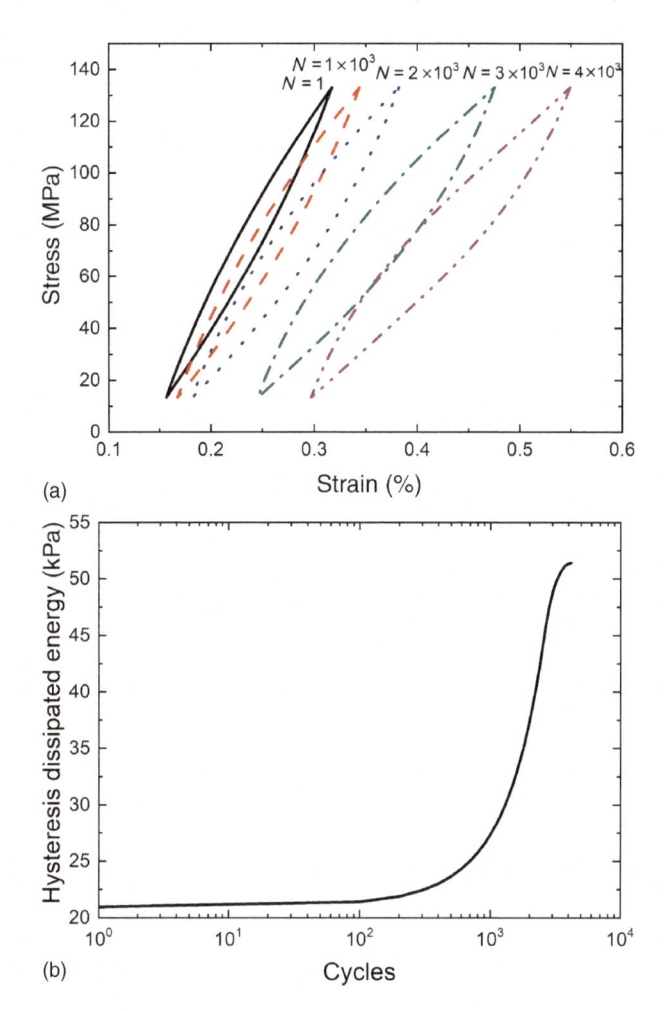

3.2 Relationships Between Interface Slip and Temperature Rising in CMCs | 125

Figure 3.9 (a) The fatigue hysteresis loops corresponding to different applied cycle number; (b) the fatigue hysteresis dissipated energy versus applied cycle number; (c) the temperature increase ($T_n - T_{initial}$) versus applied cycle number; and (d) the hysteresis dissipated energy-based damage parameter Π and temperature rise-based damage parameter Ψ versus applied cycle number of cross-ply SiC/CAS composite under fatigue peak stress of $\sigma_{max} = 133$ MPa at room temperature.

$E_f = 220$ GPa, $E_m = 350$ GPa $r_f = 3.5$ μm, $\alpha_f = -0.38 \times 10^{-6}$ °C, $\alpha_m = 4.6 \times 10^{-6}$ °C, $\Delta T = -1000$ °C, $\zeta_d = 0.2$ J/m², $\tau_{io} = 20$ MPa, $\tau_{imin} = 5$ MPa, $\omega = 6 \times 10^{-5}$, $\lambda = 1.2$, $p_1 = 0.01$, $p_2 = 1.0$, $\sigma_c = 3.5$ GPa, and $m_f = 2$.

When the fatigue peak stress is $\sigma_{max} = 335$ MPa, the stress–strain hysteresis loops corresponding to the applied cycle number of $N = 1$, 1000, 3000, 5000, and 10 000 are shown in Figure 3.10a, corresponding to the interface slip Cases I, II, III, IV, and IV, respectively. The fatigue hysteresis dissipated energy U_e versus applied cycle number curve is shown in Figure 3.10b, in which the fatigue

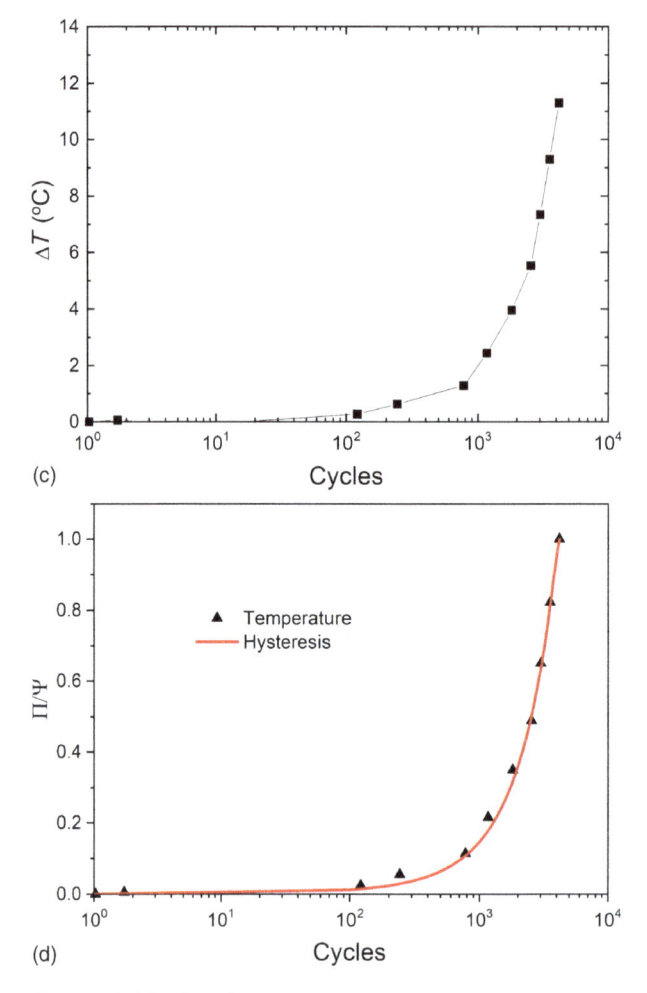

Figure 3.9 (Continued)

hysteresis dissipated energy increases from $U_e = 171.3$ kPa at $N = 1$ to the peak value of $U_e = 246.8$ kPa at $N = 3300$, then decreases to $U_e = 219$ kPa at $N = 15\,000$. The temperature increase (i.e. $T_n - T_{initial}$) versus applied cycle number curve is shown in Figure 3.10c, in which the temperature rise increase from zero at $N = 1$ to 11.7 K at $N = 15\,000$. When fatigue peak stress level is low, the majority of heating is caused by matrix multicracking, fiber/matrix interface debonding, and interface slipping. When the rate of damage accumulation decreases with applied cycles, the hysteresis heating cannot compensate the natural heat dissipation, the specimen begins to decay. The hysteresis dissipated energy-based damage parameter Π and the temperature rise-based damage parameter Ψ versus cycle number curves are given in Figure 3.10d, in which the evolution of hysteresis dissipated energy-based damage parameter Π is in coincidence with that of temperature rise-based damage parameter Ψ.

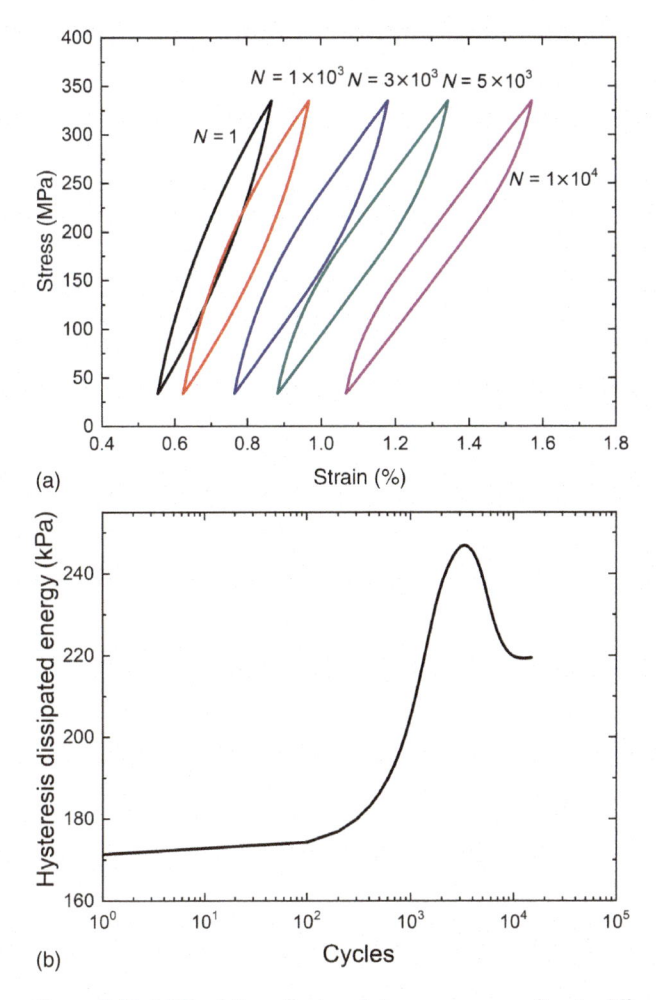

Figure 3.10 (a) The fatigue hysteresis loops corresponding to different applied cycle number; (b) the fatigue hysteresis dissipated energy versus applied cycle number; (c) the temperature increase ($T_n - T_{initial}$) versus applied cycle number; and (d) the hysteresis dissipated energy-based damage parameter Π and temperature rise-based damage parameter Ψ versus applied cycle number of 2D C/SiC composite under fatigue peak stress of $\sigma_{max} = 335$ MPa at room temperature.

3.3 Interface Assessment of CMCs from Hysteresis Loops

In this section, the fiber/matrix interface shear stress of unidirectional, cross-ply, and 2.5D woven C/SiC composites has been estimated from fatigue hysteresis loss energy at room and elevated temperatures. The experimental fatigue hysteresis loss energy and fatigue hysteresis modulus of different applied cycles have been analyzed. The theoretical fatigue hysteresis loss energy is formulated in terms of the interface shear stress. With the number of cycles increasing, the relationships

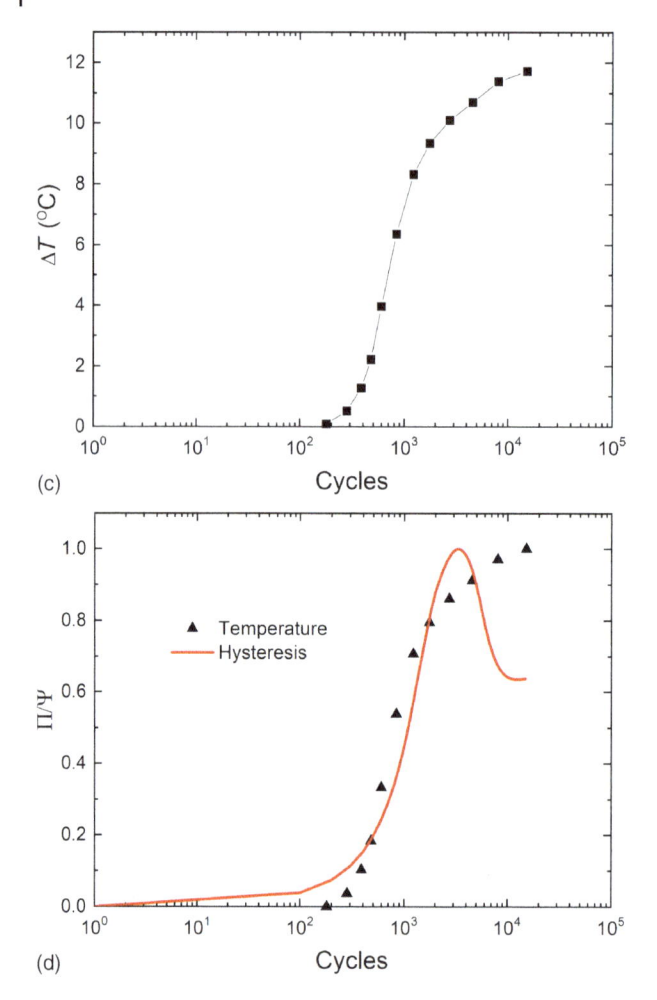

Figure 3.10 (Continued)

among the fatigue hysteresis loss energy, the fatigue hysteresis loops, the interface slip, and the interface shear stress have been established. By comparing the experimental fatigue hysteresis loss energy with theoretical computational values, the evolution of the interface shear stress versus the number of cycles has been analyzed for C/SiC composites with different fiber preforms at different experimental conditions.

3.3.1 Results and Discussion

If matrix cracking and fiber/matrix interface debonding occur upon first loading to the fatigue peak stress, the fatigue hysteresis loops develop as a result of the energy dissipation through frictional slip between fiber and matrix under fatigue loading. The shape, area, and location of fatigue hysteresis loops depend upon matrix cracking, interface debonding, interface wear at room temperature, or interface oxidation at elevated temperatures.

In cross-ply or 2.5D woven CMCs, it is assumed that the fatigue hysteresis behavior is mainly controlled by the interface frictional slip between fiber and matrix in 0° plies or longitudinal yarns. Based on fatigue hysteresis loops models developed in references, the relationships among the fatigue hysteresis loops, the fatigue hysteresis loss energy, the interface slip and the interface shear stress of unidirectional, cross-ply, and 2.5D woven C/SiC composites can be analyzed.

3.3.1.1 Unidirectional C/SiC Composite

The fatigue hysteresis loops, the fatigue hysteresis dissipated energy, and the interface slip of unidirectional C/SiC composite under tensile fatigue peak stress of $\sigma_{max} = 240$ MPa are shown in Figure 3.11. The fatigue hysteresis loops of

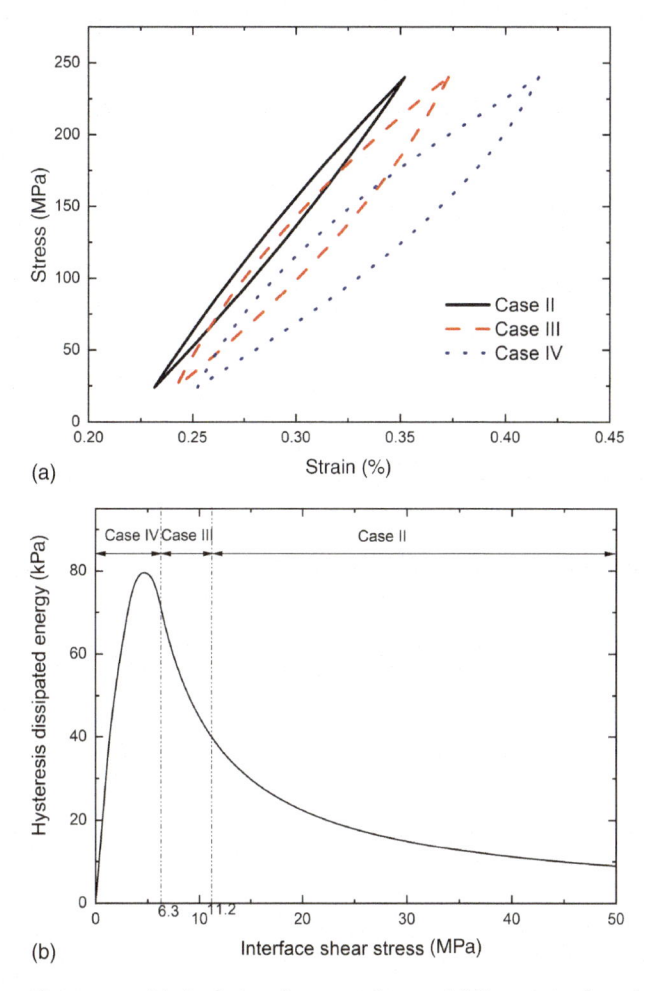

Figure 3.11 (a) The fatigue hysteresis loops of different interface slip cases; (b) the fatigue hysteresis dissipated energy versus the interface shear stress; (c) the interface debonding length ($2l_d/l_c$) versus the interface shear stress; and (d) the interface counter-slip length (y/l_d) versus the interface shear stress of unidirectional C/SiC composite under tensile fatigue peak stress of $\sigma_{max} = 240$ MPa.

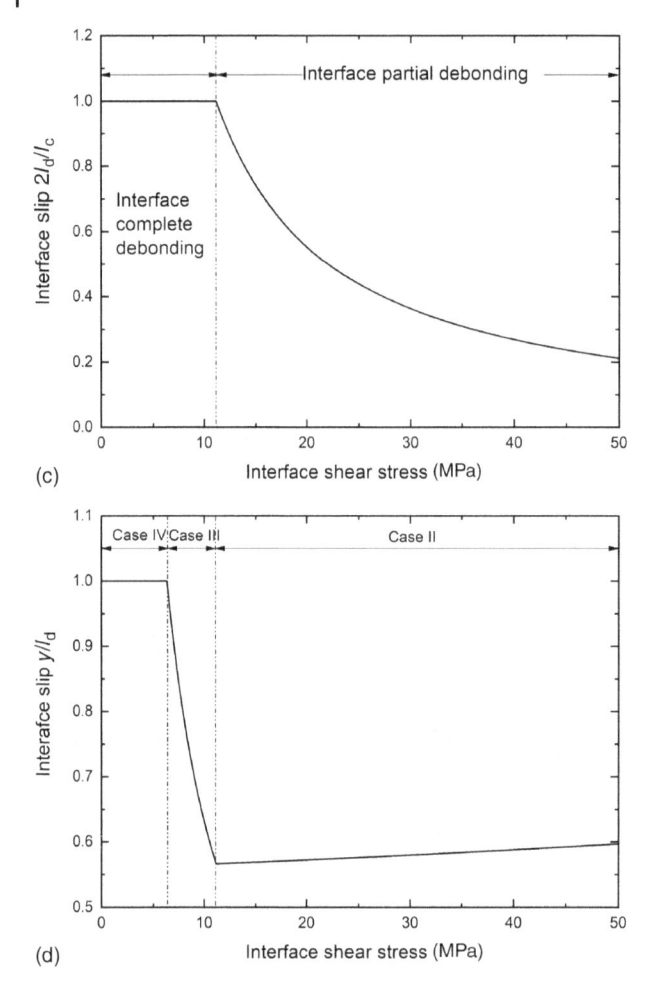

Figure 3.11 (Continued)

interface slip Cases II, III, and IV are shown in Figure 3.11a. The shape, location, and area are different for the three interface slip cases. When the interface shear stress is $\tau_i = 11.2$–50 MPa, the fatigue hysteresis dissipated energy increases with decreasing interface shear stress. The fatigue hysteresis loops correspond to the interface slip Case II, i.e. the interface partial debonding ($l_d < l_c/2$ in Figure 3.11c), and fiber sliding partial relative to the matrix in the interface debonding region ($y(\sigma_{min}) < l_d$ in Figure 3.11d) upon unloading and reloading. When the interface shear stress is $\tau_i = 6.3$–11.2 MPa, the fatigue hysteresis dissipated energy increases with decreasing interface shear stress. The fatigue hysteresis loops correspond to the interface slip Case III, i.e. the interface complete debonds ($l_d = l_c/2$ in Figure 3.11c), and fiber sliding partial relative to the matrix in the interface debonding region ($y(\sigma_{min}) < l_c/2$ in Figure 3.11d) upon unloading and reloading. When the interface shear stress is $\tau_i = 0$–6.3 MPa, the fatigue hysteresis dissipated energy increases to the peak value and then decreases with decreasing interface shear stress. The fatigue hysteresis loops correspond to the interface slip Case IV, i.e. the interface complete debonding ($l_d = l_c/2$ in Figure 3.11c), and fiber sliding complete relative to the matrix in the interface debonding region ($y(\sigma_{min}) = l_c/2$ in Figure 3.11d) upon unloading and reloading.

3.3.1.2 Cross-Ply C/SiC Composite

Kuo and Chou (1995) investigated the undamaged and damaged states of cross-ply CMCs and classified the damaged states into five matrix cracking modes, as shown in Figure 3.12. Under tensile fatigue loading of cross-ply C/SiC composite, the matrix cracking mode 3 and mode 5 both exist in the composite.

The fatigue hysteresis loops, the fatigue hysteresis dissipated energy, and the interface slip of cross-ply C/SiC composite for matrix cracking mode 3 under fatigue peak stress of $\sigma_{max} = 105$ MPa are shown in Figure 3.13. The fatigue hysteresis loops of interface slip Cases II, III, and IV are shown in Figure 3.13a. The shape, location, and area are different for the three interface slip cases. When the interface shear stress is $\tau_i = 14.7$–50 MPa, the fatigue hysteresis dissipated energy increases with decreasing interface shear stress. The fatigue hysteresis loops correspond to the interface slip Case II, i.e. the interface partial debonding

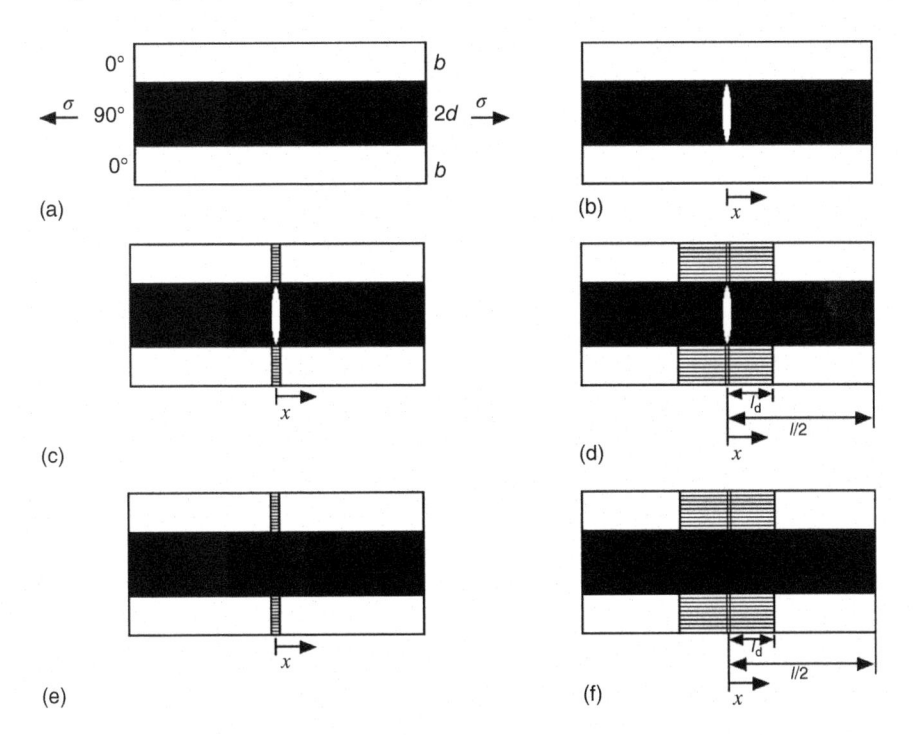

Figure 3.12 The undamaged state and five damaged modes of cross-ply ceramic composite (a) undamaged composite; (b) mode 1: transverse cracking; (c) mode 2: transverse cracking and matrix cracking with perfect fiber/matrix bonding; (d) mode 3: transverse cracking and matrix cracking with fiber/matrix interface debonding; and (e) mode 4: matrix cracking with perfect fiber/matrix bonding; (f) mode 5: matrix cracking with fiber/matrix debonding.

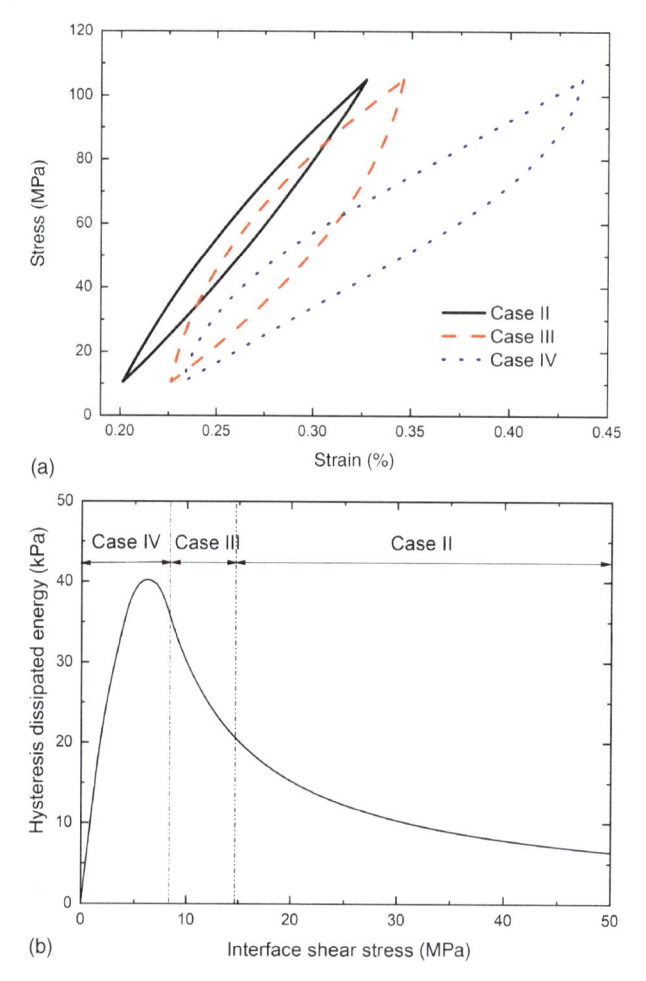

Figure 3.13 (a) The fatigue hysteresis loops of different interface cases; (b) the fatigue hysteresis dissipated energy versus the interface shear stress; (c) the interface debonded length $(2l_d/l_c)$ versus the interface shear stress; and (d) the interface counter-slip length (y/l_d) versus the interface shear stress of matrix cracking mode 3 of cross-ply C/SiC composite under tensile fatigue peak stress of $\sigma_{max} = 105$ MPa.

($l_d < l_c/2$ in Figure 3.13c), and the fiber sliding partial relative to the matrix in the interface debonding region ($y(\sigma_{min}) < l_d$ in Figure 3.13d) upon unloading and reloading. When the interface shear stress is $\tau_i = 8.3$–14.7 MPa, the fatigue hysteresis dissipated energy increases with decreasing interface shear stress. The fatigue hysteresis loops correspond to the interface slip Case III, i.e. the interface complete debonding ($l_d = l_c/2$ in Figure 3.13c), and the fiber sliding partial relative to the matrix in the interface debonding region ($y(\sigma_{min}) < l_c/2$ in Figure 3.13d) upon unloading and reloading. When the interface shear stress is $\tau_i = 0$–8.3 MPa, the fatigue hysteresis dissipated energy increases to the peak value and then decreases with decreasing interface shear stress. The fatigue hysteresis loops correspond to the interface slip Case IV, i.e. the interface complete

(c)

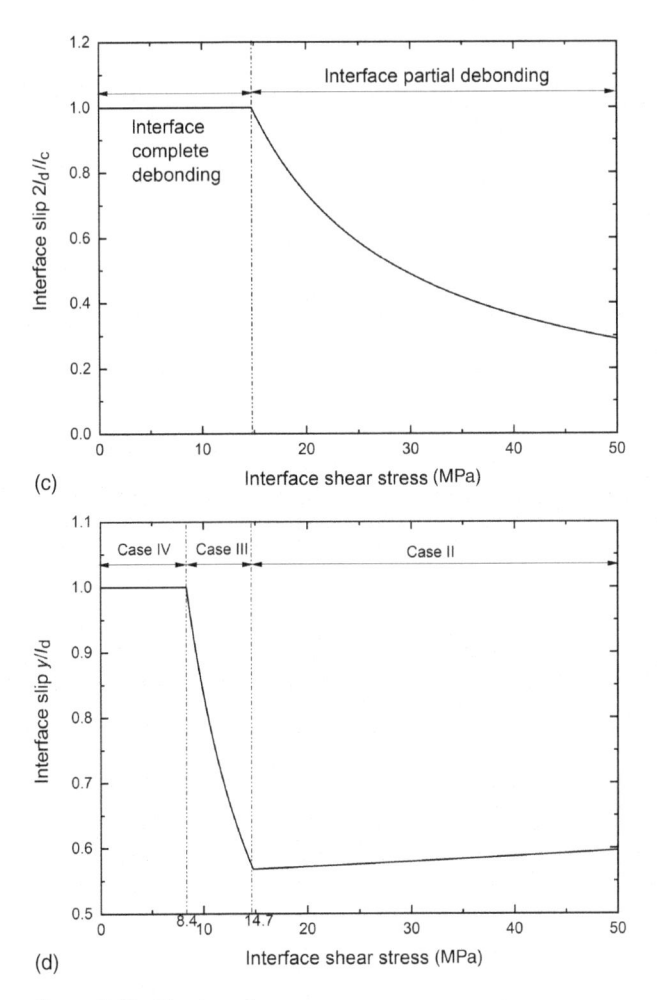

(d)

Figure 3.13 (Continued)

debonding ($l_d = l_c/2$ in Figure 3.13c), and the fiber sliding complete relative to the matrix in the interface debonding region ($y(\sigma_{min}) = l_c/2$ in Figure 3.13d) upon unloading and reloading.

The fatigue hysteresis loops, the fatigue hysteresis dissipated energy and the interface slip of cross-ply C/SiC composite for matrix cracking mode 5 under fatigue peak stress of $\sigma_{max} = 105$ MPa are shown in Figure 3.14. The fatigue hysteresis loops of interface slip Cases II, III, and IV are shown in Figure 3.14a. The shape, location, and area are different for the three interface slip cases. When the interface shear stress is $\tau_i = 13$–50 MPa, the fatigue hysteresis dissipated energy increases with decreasing interface shear stress. The fatigue hysteresis loops correspond to the interface slip Case II, i.e. the interface partial debonding ($l_d < l_c/2$ in Figure 3.14c), and fiber sliding partial relative to the matrix in the interface debonding region ($y(\sigma_{min}) < l_d$ in Figure 3.14d) upon unloading and reloading. When the interface shear stress is $\tau_i = 7.2$–13 MPa, the fatigue

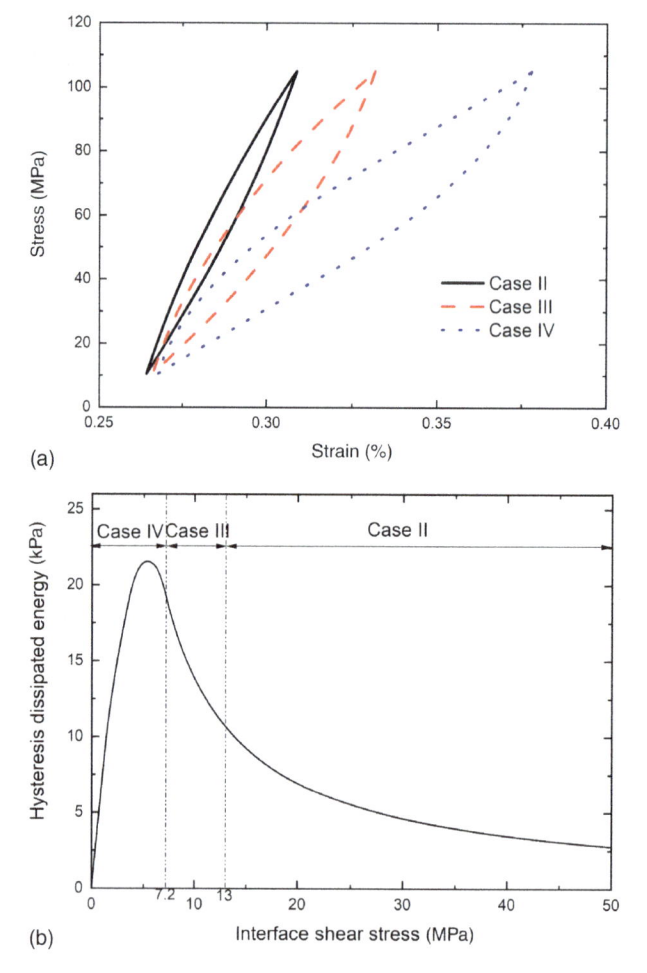

Figure 3.14 (a) The fatigue hysteresis loops of different interface cases; (b) the fatigue hysteresis dissipated energy versus the interface shear stress; (c) the interface debonded length $(2l_d/l_c)$ versus the interface shear stress; and (d) the unloading interface counter-slip length (y/l_d) versus the interface shear stress of matrix cracking mode 5 of cross-ply C/SiC composite under tensile fatigue peak stress of $\sigma_{max} = 105$ MPa.

hysteresis dissipated energy increases with decreasing interface shear stress. The fatigue hysteresis loops correspond to the interface slip Case III, i.e. the interface complete debonding ($l_d = l_c/2$ in Figure 3.14c), and the fiber sliding partial relative to the matrix in the interface debonding region ($y(\sigma_{min}) < l_c/2$ in Figure 3.14d) upon unloading and reloading. When the interface shear stress is $\tau_i = 0$–7.2 MPa, the fatigue hysteresis dissipated energy increases to the peak value and then decreases with interface shear stress. The fatigue hysteresis loops correspond to the interface slip Case IV, i.e. the interface complete debonding ($l_d = l_c/2$ in Figure 3.14c), and the fiber sliding complete relative to the matrix in the interface debonding region ($y(\sigma_{min}) = l_c/2$ in Figure 3.14d) upon unloading and reloading.

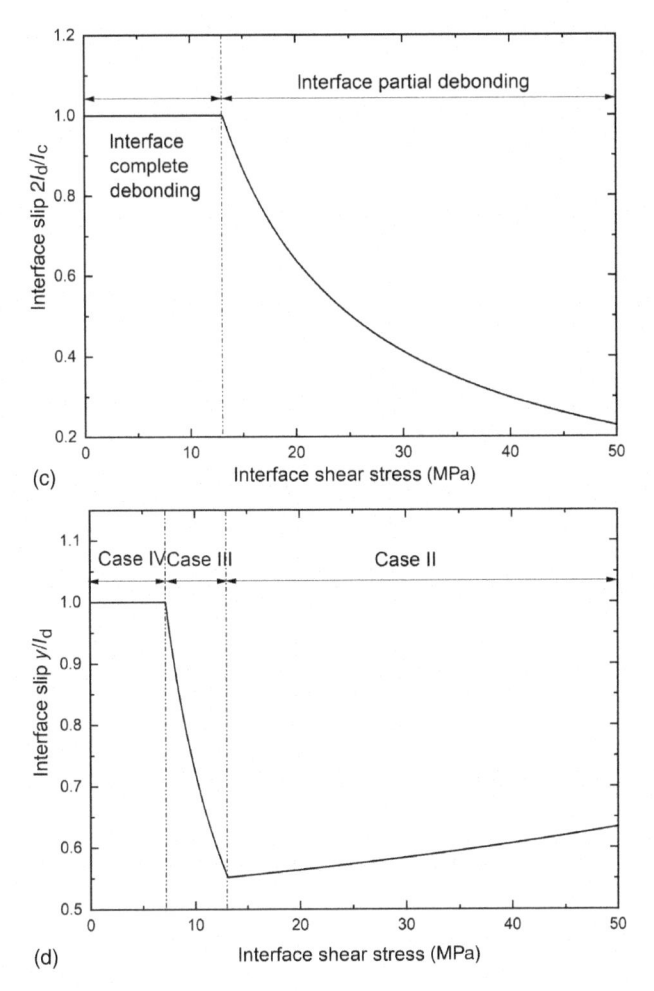

Figure 3.14 (Continued)

In Figure 3.15a, it shows that the shape, location, and area of fatigue hysteresis loops of single matrix cracking mode 3 or mode 5 exhibit differences. As matrix cracking mode 3 and mode 5 both exist in cross-ply C/SiC composite under fatigue loading, the fatigue hysteresis loops considering matrix cracking mode 3 and mode 5 together are shown in Figure 3.15a. The damage mode parameter η meaning the proportion of matrix cracking mode 3 in cross-ply C/SiC composite is approximately 0.40 at the fatigue peak stress of $\sigma_{max} = 105$ MPa. The fatigue hysteresis dissipated energy versus the interface shear stress curves for matrix cracking mode 3, mode 5, and composite are shown in Figure 3.15b, in which the curve of the composite is between curves of matrix cracking mode 3 and mode 5.

3.3.1.3 2.5D C/SiC Composite

Based on microstructure analysis of 2.5D woven CMCs, the composite can be divided into four elements of 0° warp yarns, 90° weft yarns, matrix outside of

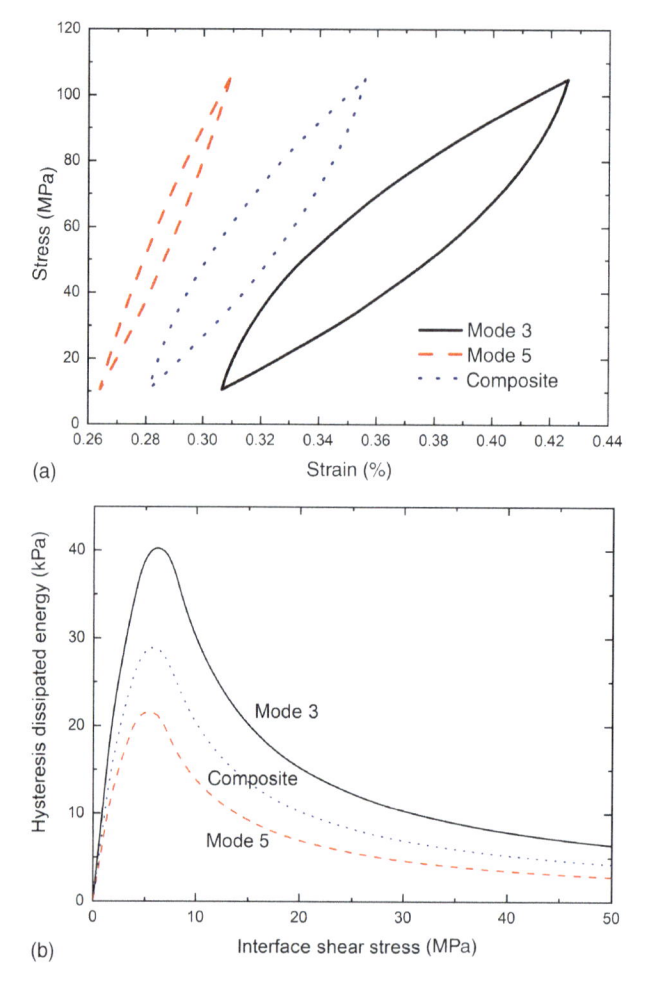

Figure 3.15 (a) The fatigue hysteresis loops of matrix cracking mode 3, mode 5, and cross-ply composite; (b) the fatigue hysteresis loss energy versus the interface shear stress curves for matrix cracking mode 3, mode 5, and cross-ply C/SiC composite.

yarns, and open porosity. When matrix cracking and interface debonding occur in 0° warp yarns upon first loading to fatigue peak stress, the fatigue hysteresis loops develop due to frictional slip between fiber and matrix in 0° warp yarns.

The fatigue hysteresis loops, the fatigue hysteresis dissipated energy, and the interface slip of 2.5D C/SiC composite under fatigue peak stress of $\sigma_{max} = 200$ MPa are shown in Figure 3.16. The fatigue hysteresis loops of 2.5D C/SiC composite for the interface slip Cases II, III, and IV are shown in Figure 3.16a. The shape, location, and area are different for the three interface slip cases. When the interface shear stress is $\tau_i = 10.5$–50 MPa, the fatigue hysteresis dissipated energy increases with decreasing interface shear stress. The fatigue hysteresis loops correspond to the interface slip Case II, i.e. the interface partial debonding ($l_d < l_c/2$ in Figure 3.16c), and the fiber sliding partial relative to the

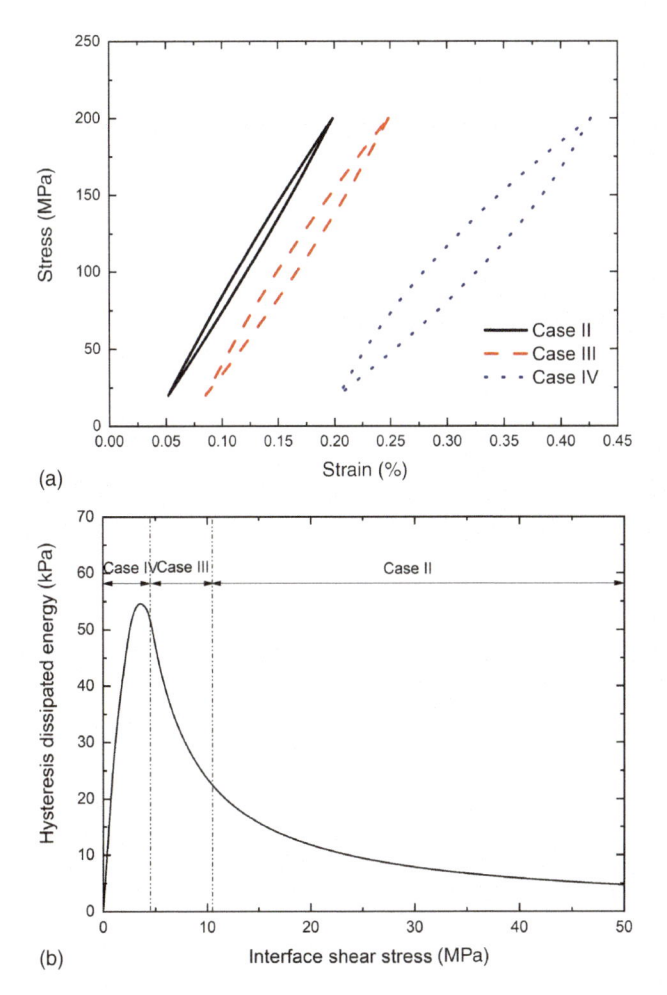

Figure 3.16 (a) The fatigue hysteresis loops of different interface cases; (b) the fatigue hysteresis dissipated energy versus the interface shear stress; (c) the interface debonding length $(2l_d/l_c)$ versus the interface shear stress; and (d) the unloading interface counter-slip length (y/l_d) versus the interface shear stress of 2.5D C/SiC composite under tensile fatigue peak stress of $\sigma_{max} = 200$ MPa.

matrix in the interface debonding region ($y(\sigma_{min}) < l_d$ in Figure 3.16d) upon unloading and reloading. When the interface shear stress is $\tau_i = 4.8$–10.5 MPa, the fatigue hysteresis dissipated energy increases with decreasing interface shear stress. The fatigue hysteresis loops correspond to the interface slip Case III, i.e. the interface complete debonding ($l_d = l_c/2$ in Figure 3.16c), and the fiber sliding partial relative to the matrix in the interface debonding region ($y(\sigma_{min}) < l_c/2$ in Figure 3.16d) upon unloading and reloading. When the interface shear stress is $\tau_i = 0$–4.8 MPa, the fatigue hysteresis dissipated energy increases to the peak value and then decreases with decreasing interface shear stress. The fatigue hysteresis loops correspond to the interface slip Case IV, i.e. the interface complete debonding ($l_d = l_c/2$ in Figure 3.16c) and the fiber sliding complete relative to

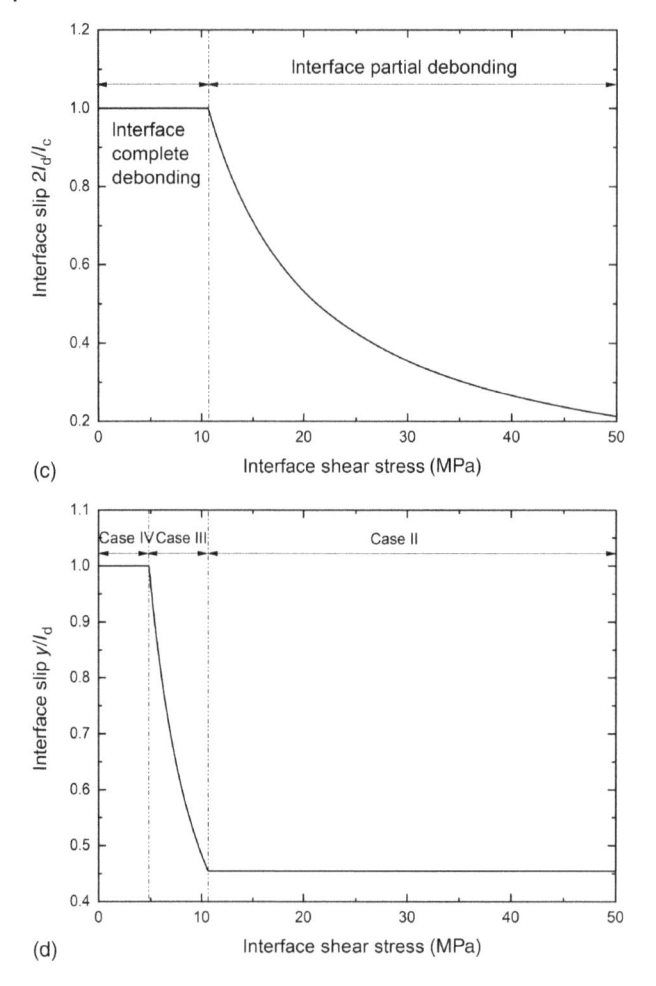

Figure 3.16 (Continued)

the matrix in the interface debonding region ($y(\sigma_{min}) = l_d$ in Figure 3.16d) upon unloading and reloading.

3.3.2 Experimental Comparisons

Comparing experimental fatigue hysteresis dissipated energy with theoretical computational values derived from fatigue hysteresis loops models of unidirectional, cross-ply, and 2.5D woven composites, the interface shear stress of CMCs with different fiber preforms can be estimated for different applied cycles at room and elevated temperatures.

3.3.2.1 Unidirectional C/SiC Composite

Under the fatigue peak stress of $\sigma_{max} = 240$ MPa, the experimental fatigue hysteresis loops of the 1st, 10th, 10 000th, 100 000th, and 1 000 000th cycles and the corresponding fatigue hysteresis dissipated energy are shown in Figure 3.17a.

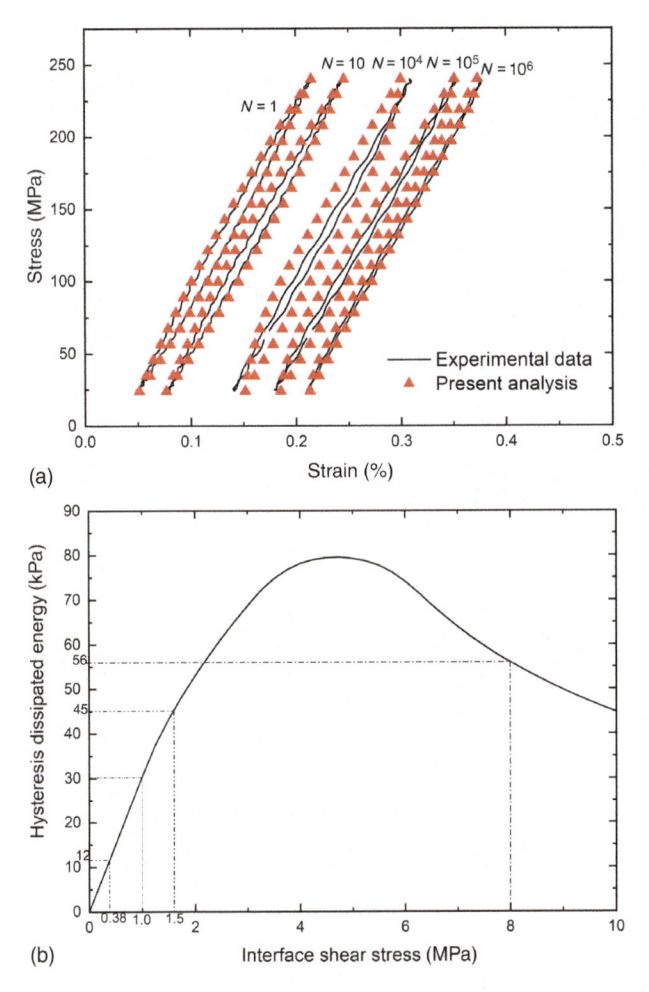

Figure 3.17 (a) The experimental and predicted fatigue hysteresis loops; (b) the theoretical fatigue hysteresis dissipated energy as a function of the interface shear stress of unidirectional C/SiC composite under tensile fatigue peak stress of $\sigma_{max} = 240$ MPa at room temperature.

The fatigue hysteresis dissipated energy as a function of interface shear stress is given in Figure 3.17b. The fatigue hysteresis dissipated energy increases with decreasing interface shear stress to the peak value of $U_e = 79.5$ kPa (the corresponding interface shear stress is $\tau_i = 4.7$ MPa), then decreases with decreasing interface shear stress to $U_e = 0$ (the corresponding interface shear stress is $\tau_i = 0$ MPa). Comparing the experimental fatigue hysteresis dissipated energy with theoretical computational values derived from unidirectional fatigue hysteresis loops models, the interface shear stress of the 1st, 3rd, 7th, 10th, 100th, 10 000th, 100 000th, and 1 000 000th cycles can be estimated, as shown in Table 3.1. The experimental fatigue hysteresis dissipated energy of the 1st, 3rd, 7th, 10th, 100th, 10 000th, 100 000th, and 1 000 000th cycles are 56, 45, 30, 12.7, 12, 10.4, 8.8, and 8.0 kPa, respectively; the corresponding interface shear stresses

Table 3.1 The interface shear stress of unidirectional C/SiC composite corresponding to different applied cycles under fatigue peak stress of $\sigma_{max} = 240$ MPa at room temperature.

Loading cycles	Experimental hysteresis dissipated energy (kPa)	Interface shear stress (MPa)
1	56	8.0
3	45	1.5
7	30	1.0
10	12.7	0.4
100	12.0	0.38
10 000	10.4	0.35
100 000	8.8	0.32
1 000 000	8.0	0.3

estimated from fatigue hysteresis dissipated energy are 8, 1.5, 1.0, 0.4, 0.38, 0.35, 0.32, and 0.3 MPa, respectively. Under the fatigue peak stress of $\sigma_{max} = 240$ MPa, the fatigue hysteresis dissipated energy of the first cycle lies in the right side of the fatigue hysteresis dissipated energy versus the interface shear stress curve. The fatigue hysteresis loops of the first cycle correspond to the interface slip Case II, i.e. the interface partial debonding and the fiber sliding partial relative to the matrix in the interface debonding region upon unloading and reloading. With the number of cycles increasing, the interface wear is the mainly reason for the interface shear stress degradation. When the interface complete debonds, the interface shear stress degrades rapidly due to the interface radial thermal residual tensile stress. The fatigue hysteresis loops of the 100th cycle correspond to the interface slip Case IV, i.e. the interface complete debonding and the fiber sliding complete relative to the matrix in the interface debonding region upon unloading and reloading. The theoretical predicted fatigue hysteresis loops of different cycles using unidirectional fatigue hysteresis loops models and estimated interface shear stress agreed with experimental results of unidirectional C/SiC composite, as shown in Figure 3.17a.

Under the fatigue peak stress of $\sigma_{max} = 250$ MPa at 800 °C in air atmosphere, the experimental fatigue hysteresis loops of the 1st, 1000th, 5000th, 10 000th, 15 000th, 20 000th, and 24 000th cycles and the corresponding fatigue hysteresis dissipated energy are shown in Figure 3.18a. The fatigue hysteresis dissipated energy as a function of the interface shear stress is given in Figure 3.18b. The fatigue hysteresis dissipated energy increases with decreasing interface shear stress to the peak value of $U_e = 90$ kPa (the corresponding interface shear stress is $\tau_i = 4.8$ MPa), then decreases with decreasing interface shear stress to $U_e = 0$ (the corresponding interface shear stress is $\tau_i = 0$ MPa). Comparing the experimental fatigue hysteresis dissipated energy with theoretical computational values derived from unidirectional fatigue hysteresis loops models, the interface shear stress of the 1st, 100th, 1000th, 5000th, 10 000th, 15 000th, 20 000th, and 24 000th cycles can be estimated, as shown in Table 3.2. The experimental

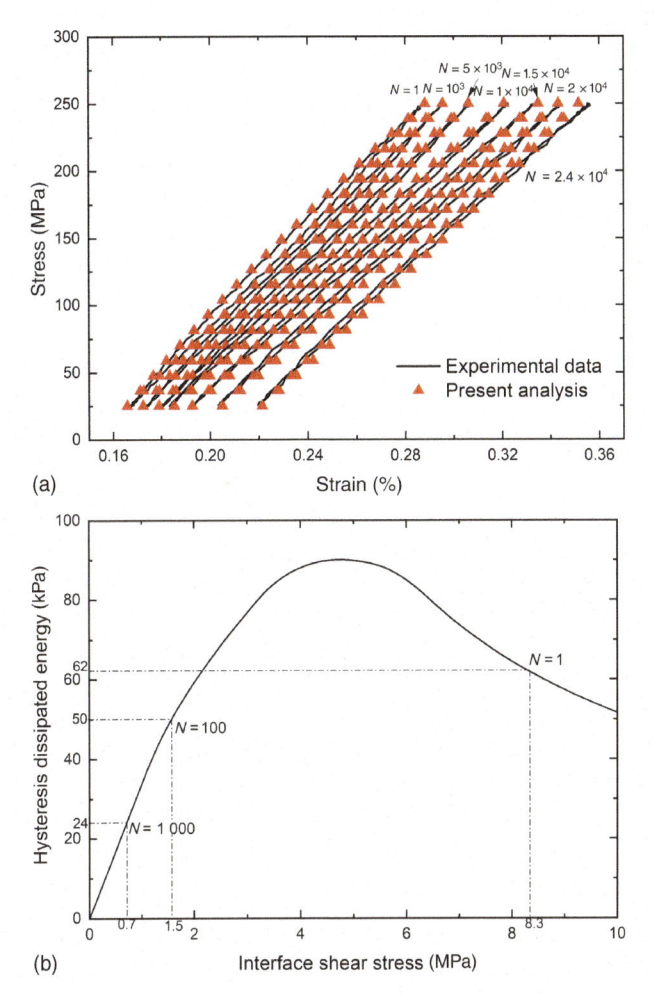

Figure 3.18 (a) The experimental and predicted fatigue hysteresis loops; (b) the theoretical fatigue hysteresis loss energy as a function of the interface shear stress of unidirectional C/SiC composite under tensile fatigue peak stress of σ_{max} = 250 MPa at an elevated temperature of 800 °C in air atmosphere.

fatigue hysteresis dissipated energy of the 1st, 100th, 1000th, 5000th, 10 000th, 15 000th, 20 000th, and 24 000th cycles are 62, 50, 24, 16, 12, 8.0, 7.8, and 7.2 kPa, respectively; the corresponding interface shear stresses estimated from fatigue hysteresis dissipated energy are 8.3, 1.5, 0.7, 0.47, 0.35, 0.24, 0.23, and 0.21 MPa, respectively. Under the fatigue peak stress of σ_{max} = 250 MPa, the fatigue hysteresis dissipated energy of the first cycle lies in the right side of the fatigue hysteresis dissipated energy versus the interface shear stress curve. The fatigue hysteresis loops of the first cycle correspond to the interface slip Case II, i.e. the interface partial debonding and the fiber sliding partial relative to the matrix in the interface debonding region upon unloading and reloading. With the number of cycles increasing, the interface oxidation is the main reason for the interface shear stress degradation. When the interface complete debonds,

Table 3.2 The interface shear stress of unidirectional C/SiC composite corresponding to different applied cycles under tensile fatigue peak stress of $\sigma_{max} = 250$ MPa at 800 °C in air atmosphere.

Loading cycles	Experimental hysteresis loss energy (kPa)	Interface shear stress (MPa)
1	62	8.3
100	50	1.5
1 000	24	0.7
5 000	16	0.47
10 000	12	0.35
15 000	8.0	0.24
20 000	7.8	0.23
24 000	7.2	0.21

the interface shear stress degrades rapidly due to the interface radial thermal residual tensile stress. The fatigue hysteresis loops of the 100th cycle correspond to the interface slip Case IV, i.e. the interface complete debonding and the fiber sliding complete relative to the matrix in the interface debonding region upon unloading and reloading. The theoretical predicted fatigue hysteresis loops of different applied cycles using unidirectional fatigue hysteresis loops models and estimated interface shear stress agreed with experimental results of unidirectional C/SiC composite, as shown in Figure 3.18a.

3.3.2.2 Unidirectional SiC/CAS Composite

Evans et al. (1995) investigated the tension–tension fatigue behavior of unidirectional SiC/CAS composite at room temperature. The fatigue loading was in a sinusoidal waveform and a frequency of $f = 10$ Hz. The tensile fatigue stress ratio was $R = 0.05$. The fatigue peak stress was $\sigma_{max} = 280$ MPa. The fatigue hysteresis loops of the 1st, 5th, 9th, and 109th cycles and the corresponding fatigue hysteresis loss energy are shown in Figure 3.19a. The fatigue hysteresis dissipated energy as a function of the interface shear stress is shown in Figure 3.19b. The fatigue hysteresis dissipated energy increases with decreasing interface shear stress to the peak value of $U_e = 81$ kPa (the corresponding interface shear stress is $\tau_i = 7$ MPa), then decreases interface shear stress to $U_e = 0$ (the corresponding interface shear stress is $\tau_i = 0$ MPa). Comparing experimental fatigue hysteresis dissipated energy with theoretical computational values derived from unidirectional fatigue hysteresis loops models, the interface shear stress of different cycles is given in Table 3.3. The experimental fatigue hysteresis dissipated energy of the 1st, 5th, 9th, and 109th cycles are 25, 55, 80, and 22 kPa, respectively; the corresponding interface shear stresses estimated from fatigue hysteresis dissipated energy are 27, 12, 8, and 1 MPa, respectively. Under the fatigue loading of $\sigma_{max} = 280$ MPa, the fatigue hysteresis dissipated energy firstly increases to the peak value, then decreases with the number of cycles increasing, which is

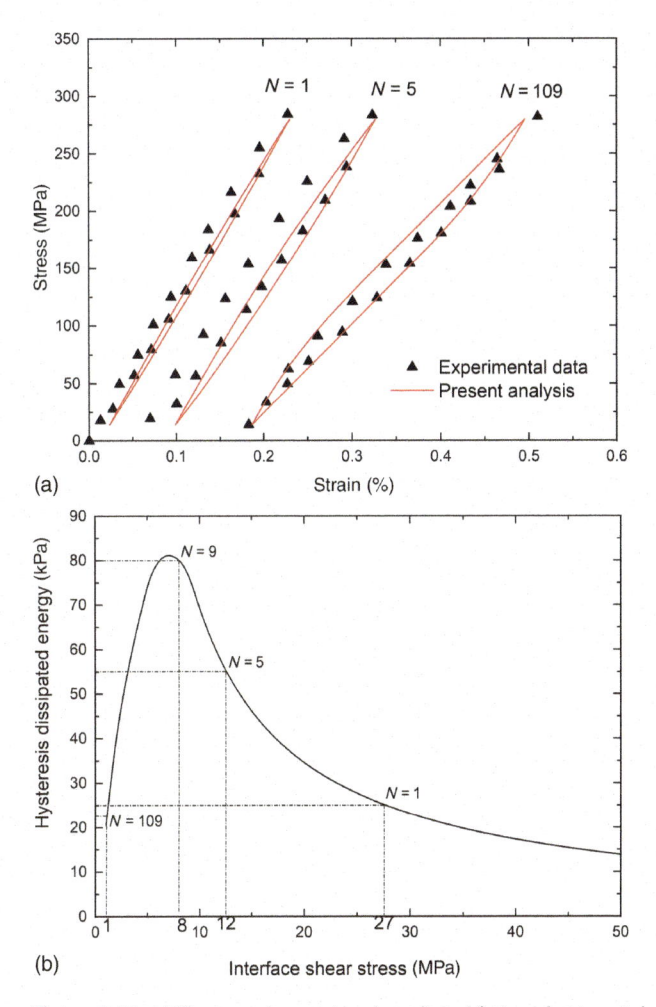

Figure 3.19 (a) The experimental and predicted fatigue hysteresis loops; (b) the theoretical fatigue hysteresis loss energy as a function of the interface shear stress of unidirectional SiC/CAS composite under tensile fatigue peak stress of $\sigma_{max} = 280$ MPa at room temperature.

Table 3.3 The interface shear stress of unidirectional SiC/CAS composite corresponding to different applied cycles under tensile fatigue peak stress of $\sigma_{max} = 280$ MPa at room temperature.

Loading cycles	Experimental hysteresis loss energy (kPa)	Interface shear stress (MPa)	Interface shear stress (MPa) (Evans et al. 1995)
1	25	27	22
5	55	12	15
9	80	8	10
109	22	1	5

corresponding to the right and left sides of the fatigue hysteresis dissipated energy versus the interface shear stress curve, respectively. The interface shear stress of different applied cycles has also been predicted by Evans's hysteresis loops measurement (Evans et al. 1995), as shown in Table 3.3, in which the interface shear stress predicted by the present analysis agreed with values given by Evans's results [0]. The theoretical predicted fatigue hysteresis loops of different cycles using unidirectional fatigue hysteresis loops models and estimated interface shear stress agreed with experimental results of unidirectional SiC/CAS composite, as shown in Figure 3.19a.

3.3.2.3 Unidirectional SiC/CAS-II Composite

Holmes and Cho (1992) investigated the tension–tension fatigue behavior of unidirectional SiC/CAS-II composite at room temperature. The fatigue loading was in a sinusoidal waveform and a frequency of $f = 25$ Hz. The fatigue peak and valley stresses were $\sigma_{max} = 180$ and $\sigma_{min} = 10$ MPa, respectively. The fatigue hysteresis loops of the 10th, 20th, and 3200th cycles and the corresponding fatigue hysteresis dissipated energy are shown in Figure 3.20a. The fatigue hysteresis dissipated energy as a function of the interface shear stress is given in Figure 3.20b. The fatigue hysteresis dissipated energy increases with decreasing interface shear stress to the peak value of $U_e = 42.6$ kPa (the corresponding interface shear stress is $\tau_i = 3.4$ MPa), then decreases with the interface shear stress to $U_e = 0$ (the corresponding interface shear stress is $\tau_i = 0$ MPa). Comparing the experimental fatigue hysteresis dissipated energy with theoretical computational values derived from unidirectional fatigue hysteresis loops models, the interface shear stress of different cycles is given in Table 3.4. The experimental fatigue hysteresis dissipated energy of the 10th, 20th, and 3200th cycles are 9, 16, and 25 kPa, respectively; the corresponding interface shear stresses estimated from fatigue hysteresis loss energy are 19, 11, and 7 MPa, respectively. Under the fatigue loading of $\sigma_{max} = 180$ MPa, the fatigue hysteresis dissipated energy increases with applied cycles, which is corresponding to the right side of the fatigue hysteresis dissipated energy versus the interface shear stress curve. The interface shear stress of different cycles has also been predicted by Holmes's frictional heating measurements (Holmes and Cho 1992), as shown in Table 3.4, in which the interface shear stress predicted by the present analysis agreed with values given by Holmes's results (Holmes and Cho 1992). The theoretical predicted fatigue hysteresis loops of different cycles using unidirectional fatigue hysteresis loops models and estimated interface shear stress agreed with experimental results of unidirectional SiC/CAS-II composite, as shown in Figure 3.20a.

3.3.2.4 Cross-Ply C/SiC Composite

Under the fatigue peak stress of $\sigma_{max} = 105$ MPa, the fatigue hysteresis loops of the 4000th, 10 000th, 100 000th, and 1 000 000th cycles and the corresponding fatigue hysteresis dissipated energy are shown in Figure 3.21a. The fatigue hysteresis dissipated energy as a function of the interface shear stress in the 0° ply is given in Figure 3.21b. The fatigue hysteresis dissipated energy increases with decreasing interface shear stress to the peak value of $U_e = 36.4$ kPa (the corresponding interface shear stress is $\tau_i = 6.1$ MPa), then decreases with the

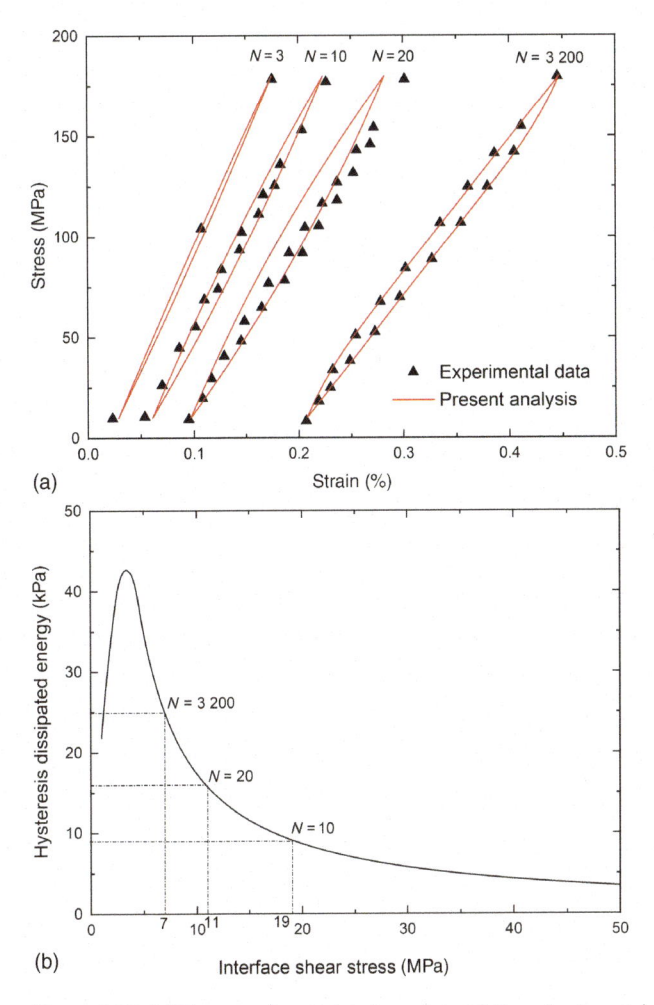

Figure 3.20 (a) The experimental and predicted fatigue hysteresis loops; (b) the theoretical fatigue hysteresis loss energy as a function of the interface shear stress of unidirectional SiC/CAS-II composite under tensile fatigue peak stress of σ_{max} = 180 MPa at room temperature.

Table 3.4 The interface shear stress of unidirectional SiC/CAS-II composite corresponding to different applied cycles under tensile fatigue peak stress of σ_{max} = 180 MPa at room temperature.

Loading cycles	Experimental hysteresis loss energy (kPa)	Interface shear stress (MPa)	Interface shear stress (MPa) (Holmes and Cho 1992)
10	9	19	20
20	16	11	8
3200	25	7	5

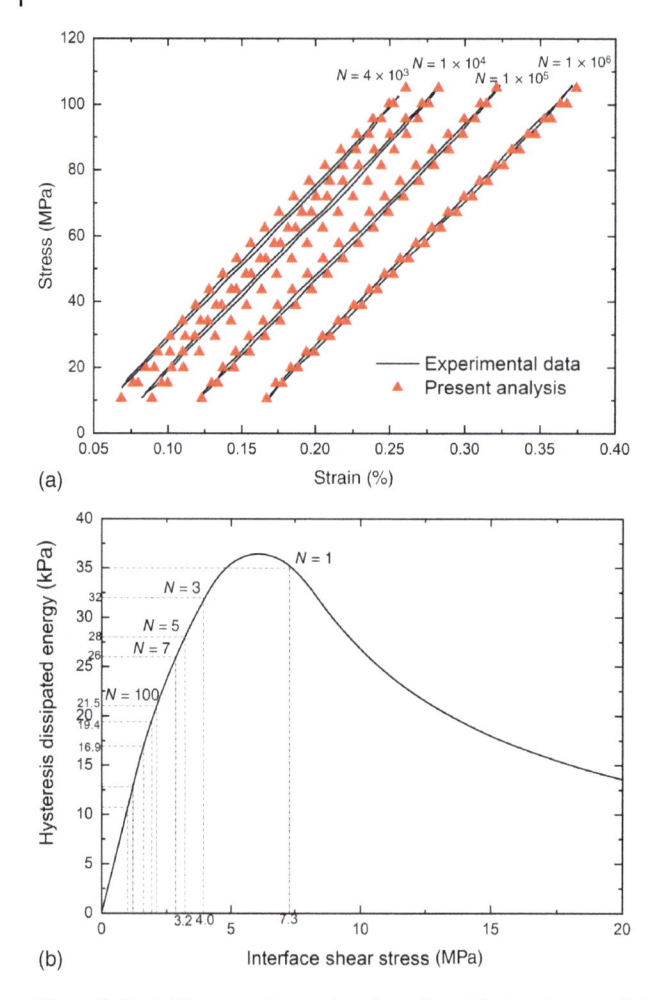

Figure 3.21 (a) The experimental and predicted fatigue hysteresis loops; (b) the theoretical fatigue hysteresis dissipated energy as a function of the interface shear stress of cross-ply C/SiC composite under tensile fatigue peak stress of $\sigma_{max} = 105$ MPa at room temperature.

interface shear stress to $U_e = 0$ (the corresponding interface shear stress is $\tau_i = 0$ MPa). Comparing experimental fatigue hysteresis dissipated energy with theoretical computational values derived from cross-ply fatigue hysteresis loops models, the interface shear stress of the 1st, 3rd, 5th, 7th, 100th, 1000th, 4000th, 10 000th, 100 000th, and 1 000 000th cycles can be estimated, as shown in Table 3.5. The experimental fatigue hysteresis dissipated energy of the 1st, 3rd, 5th, 7th, 100th, 1000th, 4000th, 10 000th, 100 000th, and 1 000 000th cycles are 35, 32, 28, 26, 21.5, 19.4, 18.2, 16.9, 12.8, and 10.7 kPa, respectively; the corresponding interface shear stresses estimated from fatigue hysteresis dissipated energy are 7.3, 4, 3.2, 2.8, 2.1, 1.9, 1.8, 1.6, 1.2, and 1 MPa, respectively. Under the fatigue peak stress of $\sigma_{max} = 105$ MPa, the fatigue hysteresis dissipated energy of the first cycle lies in the right side of the fatigue hysteresis dissipated energy versus the interface shear stress curve. The fatigue hysteresis loops of the first

Table 3.5 The interface shear stress of cross-ply C/SiC composite corresponding to different applied cycles under tensile fatigue peak stress of $\sigma_{max} = 105$ MPa at room temperature.

Loading cycles	Experimental hysteresis loss energy (kPa)	Interface shear stress (MPa)
1	35	7.3
3	32	4.0
5	28	3.2
7	26	2.8
100	21.5	2.1
1 000	19.4	1.9
4 000	18.2	1.8
10 000	16.9	1.6
100 000	12.8	1.2
1 000 000	10.7	1.0

cycle correspond to the interface slip Case II, i.e. the interface partial debonding, and the fiber sliding partial relative to the matrix in the interface debonding region of the 0° ply upon unloading and reloading. With the number of cycles increasing, the interface wear is the main reason for the interface shear stress degradation in the 0° ply. The fatigue hysteresis loops of the 100th cycle correspond to the interface slip Case IV, i.e. the interface complete debonding and the fiber sliding complete relative to the matrix in the interface debonding region of the 0° ply upon unloading and reloading. The theoretical predicted fatigue hysteresis loops of different applied cycles using cross-ply fatigue hysteresis loops models and estimated interface shear stress in the 0° ply agreed with the experimental results of cross-ply C/SiC composite, as shown in Figure 3.21a.

Under the fatigue peak stress of $\sigma_{max} = 105$ MPa at 800 °C in air atmosphere, the fatigue hysteresis loops of the 4th, 10th, 100th, 500th, 1000th, and 6000th cycles and the corresponding fatigue hysteresis dissipated energy are given in Figure 3.22a. The fatigue hysteresis dissipated energy as a function of the interface shear stress in the 0° ply is illustrated in Figure 3.22b. The fatigue hysteresis dissipated energy increases with decreasing interface shear stress to the peak value of $U_e = 25.6$ kPa (the corresponding interface shear stress is $\tau_i = 4.4$ MPa), then decreases with the interface shear stress to $U_e = 0$ (the corresponding interface shear stress is $\tau_i = 0$ MPa). Comparing the experimental fatigue hysteresis dissipated energy with theoretical computational values derived from cross-ply fatigue hysteresis loops models, the interface shear stress of the 1st, 2nd, 3rd, 4th, 10th, 100th, 500th, 1000th, 3000th, 6000th, and 6600th cycles can be estimated, as shown in Table 3.6. The experimental fatigue hysteresis dissipated energy of the 1st, 2nd, 3rd, 4th, 10th, 100th, 500th, 1000th, 3000th, 6000th, and 6600th cycles are 24.3, 20, 13, 12, 9.7, 8.6, 7.1, 6.1, 5.4, 5.2, and 5.1 kPa, respectively; the corresponding interface shear stresses estimated from fatigue hysteresis dissipated energy are 5.5, 2.3, 1.3, 1.2, 0.9, 0.8, 0.6, 0.5, 0.45, 0.43, and 0.4 MPa, respectively. Under the fatigue peak stress of $\sigma_{max} = 105$ MPa, the fatigue hysteresis dissipated energy of the first cycle lies in the right side of

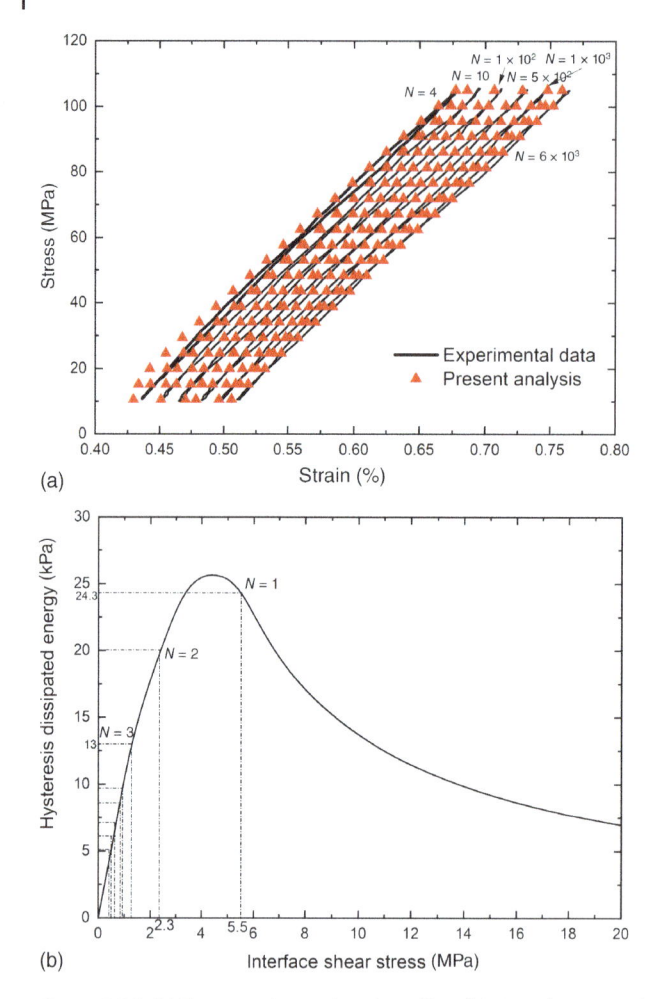

Figure 3.22 (a) The experimental and predicted fatigue hysteresis loops; (b) the theoretical fatigue hysteresis dissipated energy as a function of the interface shear stress in the 0° ply of cross-ply C/SiC composite under tensile fatigue peak stress of $\sigma_{max} = 105$ MPa at 800 °C in air atmosphere.

the fatigue hysteresis dissipated energy versus the interface shear stress curve. The fatigue hysteresis loops of the first cycle correspond to the interface slip Case II, i.e. the interface partial debonding and the fiber sliding partial relative to the matrix in the interface debonding region of the 0° ply upon unloading and reloading. With the number of cycles increasing, the interface oxidation degrades the interface shear stress in the 0° ply. The fatigue hysteresis loops of the 100th cycle correspond to the interface slip Case IV, i.e. the interface complete debonding and the fiber sliding complete relative to the matrix in the interface debonding region of the 0° ply upon unloading and reloading. The theoretical predicted fatigue hysteresis loops corresponding to different applied cycles using cross-ply fatigue hysteresis loops models and the estimated interface shear stress agreed with experimental results of cross-ply C/SiC composite, as shown in Figure 3.22a.

Table 3.6 The interface shear stress of cross-ply C/SiC composite corresponding to different applied cycles under tensile fatigue peak stress of σ_{max} = 105 MPa at 800 °C in air atmosphere.

Loading cycles	Experimental hysteresis loss energy (kPa)	Interface shear stress (MPa)
1	24.3	5.5
2	20	2.3
3	13	1.3
4	12	1.2
10	9.7	0.9
100	8.6	0.8
500	7.1	0.6
1000	6.1	0.5
3000	5.4	0.45
6000	5.2	0.43
6600	5.1	0.4

3.3.2.5 2.5D C/SiC Composite

Yang (2011) investigated the tensile fatigue behavior of 2.5D C/SiC composite at room temperature. The fatigue loading was in a sinusoidal waveform and a frequency of $f = 10$ Hz. The tensile fatigue stress ratio was $R = 0.1$. The tensile strength was approximately $\sigma_{UTS} = 225 \pm 2$ MPa. The tensile fatigue peak stresses were $\sigma_{max} = 135$ MPa (60% tensile strength), $\sigma_{max} = 157.5$ MPa (70% tensile strength), $\sigma_{max} = 168.7$ MPa (75% tensile strength), and $\sigma_{max} = 180$ MPa (80% tensile strength), respectively. The tensile fatigue life $S-N$ curve is given in Figure 3.23.

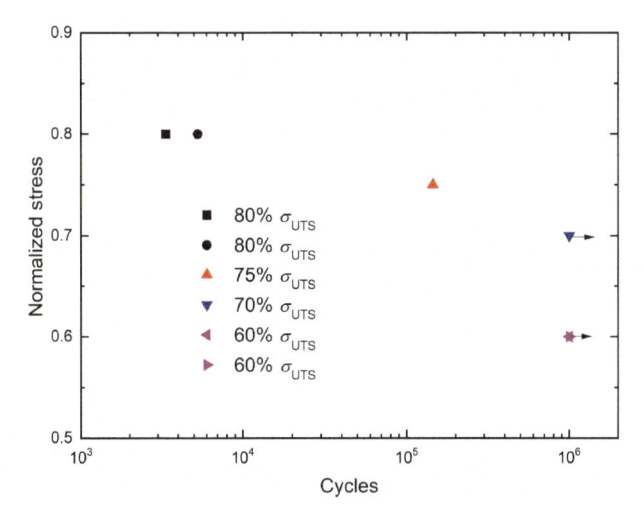

Figure 3.23 The tensile fatigue life $S-N$ curve of 2.5D C/SiC composite at room temperature.

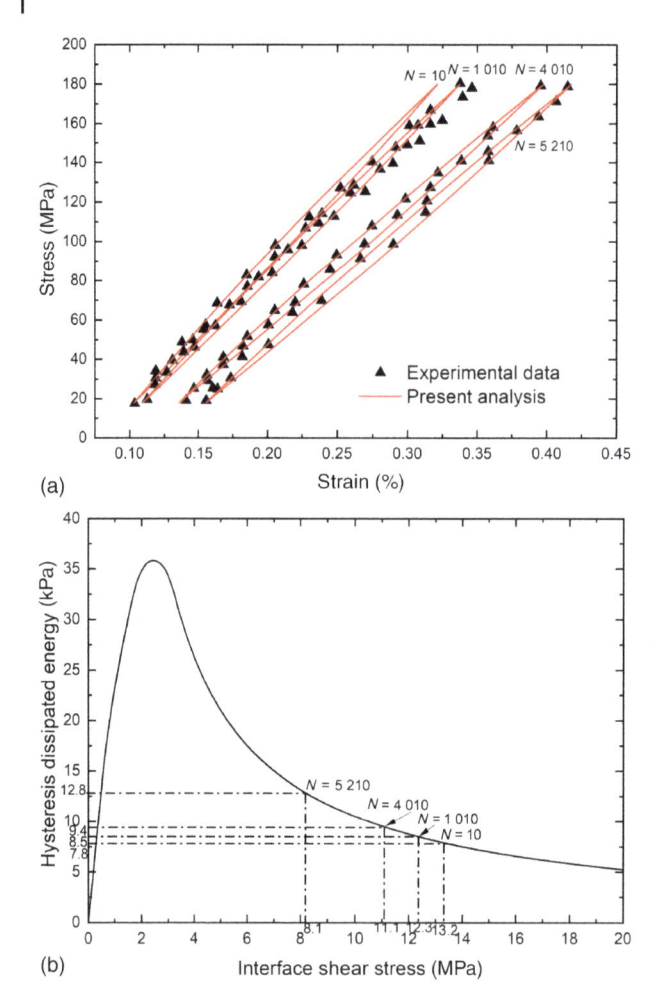

Figure 3.24 (a) The experimental and predicted fatigue hysteresis loops; (b) the theoretical fatigue hysteresis loss energy as a function of the interface shear stress of 2.5D C/SiC composite under tensile fatigue peak stress of $\sigma_{max} = 180$ MPa at room temperature.

Under the fatigue peak stress of $\sigma_{max} = 180$ MPa, the fatigue hysteresis loops of the 10th, 1010th, 4010th, and 5210th cycles and the corresponding fatigue hysteresis dissipated energy are shown in Figure 3.24a. The specimen experienced 5281 cycles and then fatigue fractured. The fatigue hysteresis dissipated energy as a function of the interface shear stress in longitudinal yarns is shown in Figure 3.24b. The fatigue hysteresis dissipated energy increases with decreasing interface shear stress to the peak value of $U_e = 35.8$ kPa (the corresponding interface shear stress is $\tau_i = 2.4$ MPa), then decreases with decrease of the interface shear stress to $U_e = 0$ (the corresponding interface shear stress is $\tau_i = 0$ MPa). Comparing the experimental fatigue hysteresis dissipated energy with theoretical computational values derived from 2.5D woven fatigue hysteresis loops models, the interface shear stress of the 10th, 1010th, 4010th, and 5210th cycles

3.3 Interface Assessment of CMCs from Hysteresis Loops

Table 3.7 The interface shear stress of 2.5D C/SiC composite corresponding to different applied cycles under tensile fatigue peak stress of σ_{max} = 180 MPa at room temperature.

Loading cycles	Experimental hysteresis loss energy (kPa)	Interface shear stress (MPa)
10	7.8	13.2
1010	8.5	12.3
4010	9.4	11.1
5210	12.8	8.1

can be estimated, as shown in Table 3.7. The experimental fatigue hysteresis dissipated energy of the 10th, 1010th, 4010th, and 5210th cycles are 7.8, 8.5, 9.4, and 12.8 kPa, respectively; the corresponding interface shear stresses estimated from fatigue hysteresis loss energy are 13.2, 12.3, 11.1, and 8.1 MPa, respectively. Under the fatigue peak stress of σ_{max} = 180 MPa, the fatigue hysteresis dissipated energy increases with the number of cycles increasing. The fatigue hysteresis loops from the 1st cycle to 5210th cycle all correspond to the interface slip Case II, i.e. the interface partial debonding and the fiber sliding partial relative to the matrix in the interface debonding region of the longitudinal yarns upon unloading and reloading. With the number of cycles increasing, the interface wear is the main reason for the interface shear stress degradation, which increases the fatigue hysteresis dissipated energy. The theoretical predicted fatigue hysteresis loops corresponding to different applied cycles using woven fatigue hysteresis loops models and estimated interface shear stress agreed with experimental results of 2.5D C/SiC composite, as shown in Figure 3.24a.

Yang (2011) investigated the tensile fatigue behavior of 2.5D C/SiC composite at 800 °C in air environment. The fatigue loading was in a sinusoidal waveform and a frequency of f = 10 Hz. The tensile fatigue stress ratio was R = 0.1. The tensile strength was approximately σ_{UTS} = 280 ± 3 MPa. The tensile fatigue peak stresses were σ_{max} = 140 MPa (50% tensile strength), 168 MPa (60% tensile strength), 196 MPa (70% tensile strength), and 224 MPa (80% tensile strength), respectively. The tensile fatigue life S–N curve is given in Figure 3.25.

Under the fatigue peak stress of σ_{max} = 140 MPa, the fatigue hysteresis loops of the 500th, 15 000th, 20 000th, and 22 700th cycles and the corresponding fatigue hysteresis dissipated energy are shown in Figure 3.26a. The specimen experienced 5281 cycles and then fatigue fractured. The fatigue hysteresis dissipated energy as a function of the interface shear stress in longitudinal yarns is shown in Figure 3.26b. The fatigue hysteresis dissipated energy increases with decreasing interface shear stress to the peak value of U_e = 21.7 kPa (the corresponding interface shear stress is τ_i = 2.25 MPa), then decreases with decreasing interface shear stress to U_e = 0 (the corresponding interface shear stress is τ_i = 0 MPa). Comparing the experimental fatigue hysteresis dissipated energy with theoretical computational values derived from 2.5D woven fatigue hysteresis loops models, the interface shear stress of the 500th, 15 000th, 20 000th, and 22 700th cycles can be estimated, as shown in Table 3.8. The experimental fatigue hysteresis dissipated

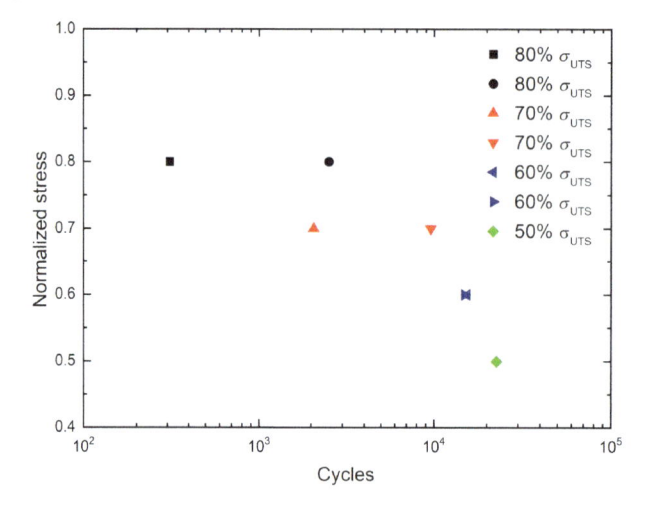

Figure 3.25 The tensile fatigue life S–N curve of 2.5D C/SiC composite at 800 °C in air atmosphere.

energy of the 500th, 15 000th, 20 000th, and 22 700th cycles are 6.3, 7.2, 8.7, and 11.8 kPa, respectively; the corresponding interface shear stresses estimated from fatigue hysteresis dissipated energy are 9.2, 8.2, 6.7, and 5.0 MPa, respectively. Under the fatigue peak stress of $\sigma_{max} = 140$ MPa, the fatigue hysteresis dissipated energy increases with the number of cycles increasing. The fatigue hysteresis loops from the 500th cycle to 22 700th cycle all correspond to the interface slip Case II, i.e. the interface partial debonding and the fiber sliding partial relative to the matrix in the interface debonding region of the longitudinal yarns upon unloading and reloading. With the number of cycles increasing, the interface oxidation degrades the interface shear stress, which increases the fatigue hysteresis dissipated energy. The theoretical predicted fatigue hysteresis loops of different cycles using woven fatigue hysteresis loops models and estimated interface shear stress agreed with experimental results of 2.5D C/SiC composite, as shown in Figure 3.26a.

Dalmaz et al. (1998) investigated the tensile fatigue behavior of 2.5D C/SiC composite at 600 °C in inert atmosphere. The fatigue peak stress was $\sigma_{max} = 230$ MPa and the valley stress was $\sigma_{min} = 0$ MPa, and the loading frequency was $f = 1$ Hz. The fatigue hysteresis loops of the 10th cycle, 10 000th cycle, and 100 000th cycle and the corresponding fatigue hysteresis dissipated energy are given in Figure 3.27a. The fatigue hysteresis dissipated energy as a function of the interface shear stress in longitudinal yarns is shown in Figure 3.27b. The fatigue hysteresis dissipated energy increases with decreasing interface shear stress to the peak value of $U_e = 108.7$ kPa (the corresponding interface shear stress is $\tau_i = 12.5$ MPa), then decreases with decrease of the interface shear stress to $U_e = 0$ (the corresponding interface shear stress is $\tau_i = 0$ MPa). Comparing the experimental fatigue hysteresis dissipated energy with theoretical computational values derived from 2.5D woven fatigue hysteresis loops models, the interface shear stress of the 10th, 10 000th, and 100 000th cycles can be estimated, as

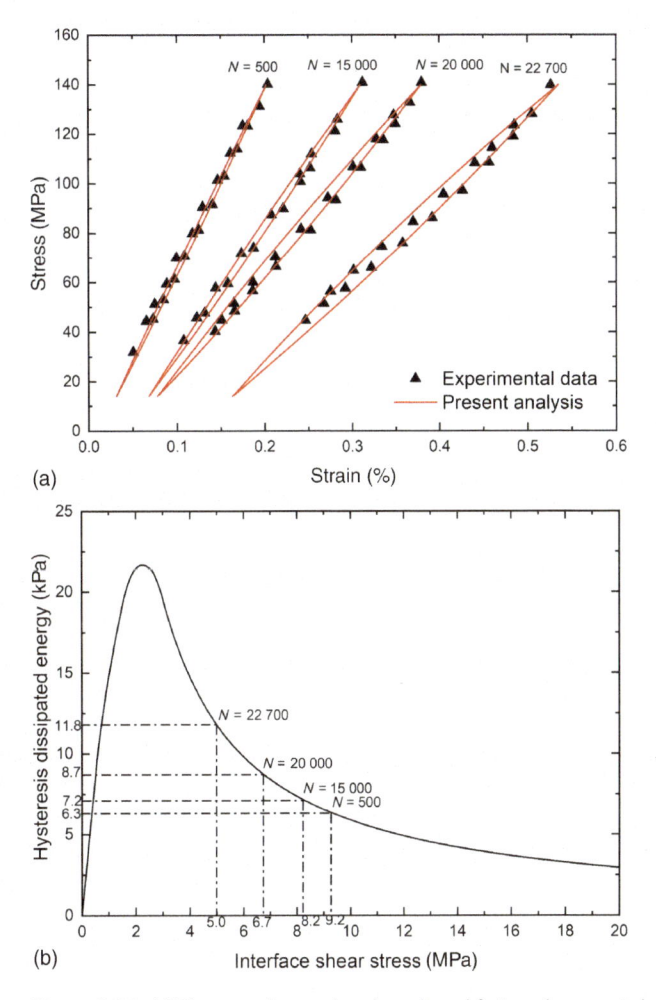

Figure 3.26 (a) The experimental and predicted fatigue hysteresis loops; (b) the theoretical fatigue hysteresis loss energy as a function of the interface shear stress of 2.5D C/SiC composite under tensile fatigue peak stress of $\sigma_{max} = 140$ MPa at 800 °C in air atmosphere.

Table 3.8 The interface shear stress of 2.5D C/SiC composite corresponding to different applied cycles under tensile fatigue peak stress of $\sigma_{max} = 140$ MPa at 800 °C in air atmosphere.

Loading cycles	Experimental hysteresis loss energy (kPa)	Interface shear stress (MPa)
500	6.3	9.2
15 000	7.2	8.2
20 000	8.7	6.7
22 700	11.8	5.0

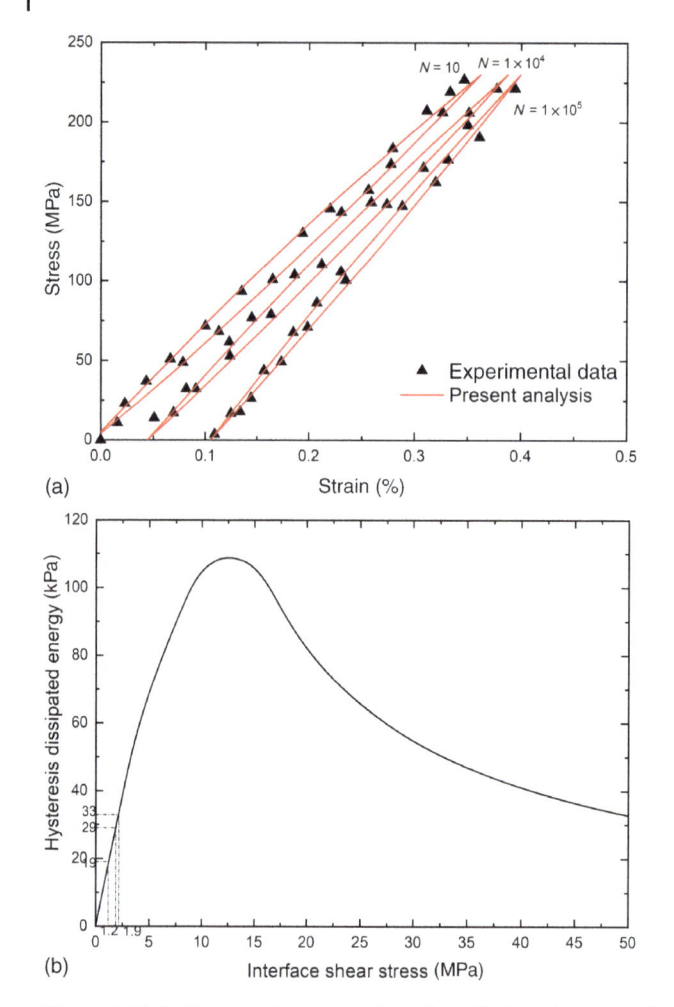

Figure 3.27 (a) The experimental and predicted fatigue hysteresis loops; (b) the theoretical fatigue hysteresis dissipated energy as a function of the interface shear stress of 2.5D C/SiC composite under tensile fatigue peak stress of $\sigma_{max} = 230$ MPa at 600 °C in inert atmosphere.

shown in Table 3.9. The experimental fatigue hysteresis dissipated energy of the 10th, 10 000th, and 100 000th cycles are 33, 29, and 19 kPa, respectively; the corresponding interface shear stresses estimated from fatigue hysteresis dissipated energy are 2.2, 1.9, and 1.2 MPa, respectively. Under the fatigue peak stress of $\sigma_{max} = 230$ MPa, the fatigue hysteresis dissipated energy decreases with the number of cycles increasing. The fatigue hysteresis loops from the 10th cycle to 100 000th cycle all correspond to the interface slip Case IV, i.e. the interface complete debonding, and the fiber sliding complete relative to the matrix in the interface debonding region of the longitudinal yarns upon unloading and reloading. With the number of cycles increasing, the interface wear degrades the interface shear stress, which degrades the fatigue hysteresis dissipated energy. The theoretical predicted fatigue hysteresis loops of different cycles using woven

Table 3.9 The interface shear stress of 2.5D C/SiC composite corresponding to different applied cycles under tensile fatigue peak stress of $\sigma_{max} = 230$ MPa at 600 °C in inert atmosphere.

Loading cycles	Experimental hysteresis loss energy (kPa)	Interface shear stress (MPa)
10	33	2.2
10 000	29	1.9
100 000	19	1.2

fatigue hysteresis loops models and estimated interface shear stress agreed with experimental results of 2.5D C/SiC composite, as shown in Figure 3.27a.

3.4 Conclusion

In this chapter, the relationships between the hysteresis dissipated energy and temperature rising of the external surface in fiber-reinforced CMCs under cyclic loading are analyzed. Based on the fatigue hysteresis theories considering fiber failure, the hysteresis dissipated energy and a hysteresis dissipated energy-based damage parameter changing with the increase of cycle number are investigated. The experimental temperature rise-based damage parameter of unidirectional SiC/CAS-II, cross-ply SiC/CAS, and 2D C/SiC composites corresponding to different fatigue peak stresses and cycle numbers are predicted. The fatigue hysteresis behavior of unidirectional, cross-ply, and 2.5D C/SiC composites at room temperature and 800 °C in air atmosphere are investigated. Comparing experimental fatigue hysteresis dissipated energy with theoretical computational values, the evolution of the interface shear stress with the number of cycles increasing is estimated.

For different fiber preforms at different test conditions, the interface shear stress of unidirectional C/SiC composite at 800 °C in air atmosphere degrades the fastest due to the low initial interface shear stress, the interface radial thermal residual tensile stress, and the interface oxidation. For 2.5D C/SiC composite, the interface shear stress at room temperature degrades the slowest due to the high initial interface shear stress, which leads to the interface partially debonding during the process of entire fatigue loading; when the number of cycles increasing, only the interface wear attributes to the interface shear stress degradation. For cross-ply C/SiC composite at the same fatigue peak stress, the interface shear stress decreases much faster at an elevated temperature of 800 °C in air atmosphere than that at room temperature.

For different fiber preforms at the same test conditions, the interface shear stress degrades the fastest for unidirectional C/SiC composite; the slowest for 2.5D C/SiC composite. However, the interface shear stress degradation rate of cross-ply C/SiC composite is between that of unidirectional and 2.5D C/SiC composites, which indicates that the interface wear or interface oxidation is much more serious in the interface debonded region of unidirectional C/SiC composite, compared with that in the interface debonded region of the 0° plies or longitudinal yarns of cross-ply or 2.5D C/SiC composites.

References

Ahn, B.K. and Curtin, W.A. (1997). Strain and hysteresis by stochastic matrix cracking in ceramic matrix composites. *Journal of the Mechanics and Physics of Solids* 45: 177–209. https://doi.org/10.1016/S0022-5096(96)00081-6.

Cho, C.D., Holmes, J.W., and Barber, J.R. (1991). Estimate of interfacial shear in ceramic composites from frictional heating measurements. *Journal of the American Ceramic Society* 74: 2802–2808. https://doi.org/10.1111/j.1151-2916.1991.tb06846.x.

Dalmaz, A., Reynaud, P., Rouby, D. et al. (1998). Mechanical behavior and damage development during cyclic fatigue at high-temperature of a 2.5D carbon/SiCcomposite. *Composites Science and Technology* 58: 693–699. https://doi.org/10.1016/S0266-3538(97)00150-4.

Dassios, K.G., Aggelis, D.G., Kordatos, E.Z., and Matikas, T.E. (2013). Cyclic loading of a SiC-fiber reinforced ceramic matrix composite reveals damage mechanisms and thermal residual stress state. *Composites Part A Applied Science and Manufacturing* 44: 105–113. https://doi.org/10.1016/j.compositesa.2012.06.011.

Domergue, J., Vagaggini, J.M., and Evans, A.G. (1995). Relationships between hysteresis measurements and the constituent properties of ceramic matrix composites: II, Experimental studies on unidirectional materials. *Journal of the American Ceramic Society* 78: 2721–2731. https://doi.org/10.1111/j.1151-2916.1995.tb08047.x.

Domergue, J.M., Heredia, F.E., and Evans, A.G. (1996). Hysteresis loops and the inelastic deformation of 0/90 ceramic matrix composites. *Journal of the American Ceramic Society* 79: 161–170. https://doi.org/10.1111/j.1151-2916.1996.tb07894.x.

Evans, A.G., Zok, F.W., and McMeeking, R.M. (1995). Fatigue of ceramic matrix composites. *Acta Metallurgica et Materialia* 43: 859–875. https://doi.org/10.1016/0956-7151(94)00304-Z.

Fantozzi, G. and Reynaud, P. (2009). Mechanical hysteresis in ceramic matrix composites. *Materials Science and Engineering A* 521–522: 18–23. https://doi.org/10.1016/j.msea.2008.09.128.

Holmes, J.W. and Cho, C.D. (1992). Experimental observation of frictional heating in fiber-reinforced ceramics. *Journal of the American Ceramic Society* 75: 929–938. https://doi.org/10.1111/j.1151-2916.1992.tb04162.x.

Holmes, J.W. and Sørensen, B.F. (1995). Fatigue behavior of continuous fiber-reinforced ceramic matrix composites. *High Temperature Mechanical Behavior of Ceramic Composites* (eds. S.V. Nair and K. Jakus), pp. 261–326. Butterworth-Heinemann.

Holmes, J.W., Wu, X., and Sorensen, B.F. (1994). Frequency dependence of fatigue life and internal heating of a fiber-reinforced/ceramic-matrix composite.

Journal of the American Ceramic Society 77: 3284–3286. https://doi.org/10.1111/j.1151-2916.1994.tb04587.x.

Hutchinson, J.W. and Jensen, H.M. (1990). Models of fiber debonding and pullout in brittle composites with friction. *Mechanics of Materials* 9: 139–163. https://doi.org/10.1016/0167-6636(90)90037-G.

Kim, J. and Liaw, P.K. (2005). Characterization of fatigue damage modes in nicalon/calcium aluminosilicate composites. *Journal of Engineering Materials and Technology* 127: 8–15. https://doi.org/10.1115/1.1836766.

Kotil, T., Holmes, J.W., and Comninou, M. (1990). Origin of hysteresis observed during fatigue of ceramic-matrix composites. *Journal of the American Ceramic Society* 73: 1879–1883. https://doi.org/10.1016/j.msea.2005.08.204.

Krenkel, W. and Berndt, F. (2005). C/C–SiC composites for space applications and advanced friction systems. *Materials Science and Engineering A* 412: 177–181. https://doi.org/10.1016/j.msea.2005.08.204.

Kuo, W.S. and Chou, T.W. (1995). Multiple cracking of unidirectional and cross-ply ceramic matrix composites. *Journal of the American Ceramic Society* 78: 745–755. https://doi.org/10.1111/j.1151-2916.1995.tb08242.x.

Li, L. (2013a). Fatigue hysteresis behavior of cross-ply C/SiC ceramic matrix composites at room and elevated temperatures. *Materials Science and Engineering A* 586: 160–170. https://doi.org/10.1016/j.msea.2013.08.017.

Li, L. (2013b). Modeling hysteresis behavior of cross-ply C/SiC ceramic matrix composites. *Composites Part B Engineering* 53: 36–45. https://doi.org/10.1016/j.compositesb.2013.04.029.

Li, L. (2014). Assessment of the interfacial properties from fatigue hysteresis loss energy in ceramic-matrix composites with different fiber preforms at room and elevated temperatures. *Materials Science and Engineering A* 613: 17–36. https://doi.org/10.1016/j.msea.2014.06.092.

Li, L. (2016). Relationship between hysteresis dissipated energy and temperature rising in fiber-reinforced ceramic-matrix composites under cyclic loading. *Applied Composite Materials* 23: 337–355. https://doi.org/10.1007/s10443-015-9463-2.

Li, L. and Song, Y. (2010). An approach to estimate interface shear stress of ceramic matrix composites from hysteresis loops. *Applied Composite Materials* 17: 309–328. https://doi.org/10.1007/s10443-009-9122-6.

Li, L. and Song, Y. (2013). Estimate interface shear stress of woven ceramic matrix composites. *Applied Composite Materials* 20: 993–1005. https://doi.org/10.1007/s10443-013-9314-y.

Liu, C.D., Cheng, L.F., Luan, X.G. et al. (2008). Damage evolution and real-time non-destructive evaluation of 2D carbon-fiber/SiC-matrix composites under fatigue loading. *Materials Letters* 62: 3922–3924. https://doi.org/10.1016/j.matlet.2008.04.063.

Mall, S. and Engesser, J.M. (2006). Effects of frequency on fatigue behavior of CVI C/SiC at elevated temperature. *Composites Science and Technology* 66: 863–874. https://doi.org/10.1016/j.compscitech.2005.06.020.

Mei, H. and Cheng, L.F. (2009). Comparison of the mechanical hysteresis of carbon/ceramic-matrix composites with different fiber preforms. *Carbon* 47: 1034–1042. https://doi.org/10.1016/j.carbon.2008.12.025.

Moevus, M., Reynaud, P., R'Mili, M. et al. (2006). Static fatigue of a 2.5D SiC/[Si–B–C] composite at intermediate temperature under air. *Advances in Science and Technology* 50: 141–146. https://doi.org/10.4028/www.scientific.net/AST.50.141.

Naslain, R.R. (2004). Design, preparation and properties of non-oxide CMCs for application in engines and nuclear reactors: an overview. *Composites Science and Technology* 64: 155–170. https://doi.org/10.1016/S0266-3538(03)00230-6.

Naslain, R.R. (2005). SiC-matrix composites: nonbrittle ceramics for thermo-structural application. *International Journal of Applied Ceramic Technology* 2: 75–84. https://doi.org/10.1111/j.1744-7402.2005.02009.x.

Pryce, A.W. and Smith, P.A. (1993). Matrix cracking in unidirectional ceramic matrix composites under quasi-static and cyclic loading. *Acta Metallurgica et Materialia* 41: 1269–1281. https://doi.org/10.1016/0956-7151(93)90178-U.

Reynaud, P. (1996). Cyclic fatigue of ceramic-matrix composites at ambient and elevated temperatures. *Composites Science and Technology* 56: 809–814. https://doi.org/10.1016/0266-3538(96)00025-5.

Reynaud, P., Rouby, D., and Fantozzi, G. (1998). Effects of temperature and of oxidation on the interfacial shear stress between fibres and matrix in ceramic-matrix composites. *Acta Materialia* 46: 2461–2469. https://doi.org/10.1016/S1359-6454(98)80029-3.

Rouby, D. and Louet, N. (2002). The frictional interface: a tribological approach of thermal misfit, surface roughness and sliding velocity effects. *Composites Part A Applied Science and Manufacturing* 33: 1453–1459. https://doi.org/10.1016/S1359-835X(02)00145-8.

Rouby, D. and Reynaud, P. (1993). Fatigue behavior related to interface modification during load cycling in ceramic-matrix fiber composites. *Composites Science and Technology* 48: 109–118. https://doi.org/10.1016/0266-3538(93)90126-2.

Ruggles-Wrenn, M.B., Delapasse, J., Chamberlain, A.L. et al. (2012). Fatigue behavior of a Hi-Nicalon™/SiC–B4C composite at 1200 °C in air and in steam. *Materials Science and Engineering A* 534: 119–128. https://doi.org/10.1016/j.msea.2011.11.049.

Shuler, S.F., Holmes, J.W., and Wu, X. (1993). Influence of loading frequency on the room-temperature fatigue of a carbon-fiber/SiC-matrix composite. *Journal of the American Ceramic Society* 76: 2327–2336. https://doi.org/10.1111/j.1151-2916.1993.tb07772.x.

Solti, J.P., Mall, S., and Robertson, D.D. (1995). Modeling behavior of cross-ply ceramic matrix composite under quasi-static loading. *Applied Composite Materials* 2: 265–292. https://doi.org/10.1007/BF00568765.

Solti, J.P., Robertson, D.D., and Mall, S. (2000). Estimation of interfacial properties from hysteresis energy loss in unidirectional ceramic matrix composites. *Advanced Composite Materials* 9: 161–173. https://doi.org/10.1163/15685510051033322.

Staehler, J.M., Mall, S., and Zawada, L.P. (2003). Frequency dependence of high-cycle fatigue behavior of CVI C/SiC at room temperature. *Composites Science and Technology* 63: 2121–2123. https://doi.org/10.1016/S0266-3538(03)00190-8.

Vagaggini, E., Domergue, J.M., and Evans, A.G. (1995). Relationships between hysteresis measurements and the constituent properties of ceramic matrix composites: I, Theory. *Journal of the American Ceramic Society* 78: 2709–2720. https://doi.org/10.1111/j.1151-2916.1995.tb08046.x.

Valentine, P.G., Rivers, H.K., and Chen, V.L. (2003). *X-37 C-SiC CMC Control Surface Components Development Status of the NASA Boeing USAF Orbital Vehicle and Related Efforts*. Langley Research Center, Marshall Space Flight Center.

Yang, F. (2011). Research on fatigue behavior of 2.5D woven ceramic matrix composites. Master thesis. Nanjing University of Aeronautics and Astronautics, Nanjing.

Yang, C.P., Jiao, G.Q., Wang, B., and Du, L. (2009). Oxidation damages and a stiffness model for 2D-C/SiC composites. *Acta Materiae Compositae Sinica* 26: 175–181.

4

Interface Damage Law of Ceramic-Matrix Composites

4.1 Introduction

Ceramic materials possess high strength and modulus at elevated temperatures. But their use as structural components is severely limited because of their brittleness. Continuous fiber-reinforced ceramic-matrix composites (CMCs), by incorporating fibers in ceramic matrices, however, can be made as strong as metal, yet are much lighter and can withstand much higher temperatures exceeding the capability of current nickel alloys used in high-pressure turbines, which can lower the fuel burn and emissions, while increasing the efficiency of aero engine (Naslain 2004; Li 2014, 2017). CMC durability has been validated through the ground testing or commercial flight testing in the demonstrator or customer gas turbine engines accumulating almost 30 000 hours of operation. The CMC combustion chamber and high-pressure turbine components were designed and tested in the ground testing of GEnx aero engine (Bednarcyk et al. 2015). The CMC rotating low-pressure turbine blades in an F414 turbofan demonstrator engine were successfully tested for 500 grueling cycles to validate the unprecedented temperature and durability capabilities by GE Aviation. The CMC tail nozzles were designed and fabricated by Snecma (SAFRAN) and completed the first commercial flight on CFM56-5B aero engine on 2015. CMCs will play a key role in the performance of CFM's LEAP turbofan engine, which would enter into service in 2016 for Airbus A320 and 2017 for Boeing 737 max.

Upon first loading to fatigue peak stress, matrix multicracking and fiber/matrix interface debonding occur (Curtin 2000). The fiber/matrix interface shear stress transfers loads between fibers and the matrix, which is critical for the inelastic behavior of CMCs. Under cyclic fatigue loading, interface wear is the dominant fatigue mechanism (Rouby and Reynaud 1993; Evans et al. 1995). The slip displacements between fibers and the matrix could reduce interface shear stress (Zhu et al. 1999). Evidences of interface wear that a reduction in the height of asperities occurs along the fiber coating for different thermal misfits, surface roughness, and frictional sliding velocity have been presented by push-out and push-back tests on a ceramic composite system (Rouby and Louet 2002). The interface wear process can be facilitated by temperature rising that occurs along the fiber/matrix interface, as frictional dissipation proceeds (Holmes and Cho 1992; Kim and Liaw 2005; Liu et al. 2008), i.e. the temperature rising exceeded 100 K under cyclic fatigue loading at 75 Hz between stress levels of 220 and 10 MPa in unidirectional SiC/CAS-II composite (Holmes and Cho 1992). Under cyclic fatigue loading at elevated temperature in air, the interphase would react to form CO if the fiber coating is carbon or PyC, resulting in a large reduction in interface shear stress. Evidences of interface oxidation, i.e. a uniform reduction in fiber diameter and a longer fiber pullout length occurs in a 2D C/SiC composite, have been presented by a non-stress oxidation experiment at 700 °C in air condition (Yang et al. 2009), and a tensile fatigue experiment at 550 °C in air condition (Mall and Engesser 2006). Moevus et al. (2006) investigated the static fatigue behavior of 2.5D C/[Si-B-C] composite at 1200 °C in air. The hysteresis loops area after a static fatigue of 144 hours under a steady stress of 170 MPa, significantly decreased, attributed to a decrease of interface shear stress caused by PyC interphase recession by oxidation. There are currently several approaches used to determine fiber/matrix interface shear stress, i.e. fiber pullout (Brandstetter et al. 2005), fiber push-in (Kuntz and Grathwahl 2001) and push-out (Chandra and Ghonem 2001), and etc. However, these approaches can only get individual fiber's interfacial properties at room temperature and only provide information regarding the interface shear stress, which would exist under monotonic loading conditions.

Under cyclic fatigue loading, the hysteresis loops appear as the fiber slips relative to matrix in the interface debonded region (Reynaud 1996). The shape, location, and area of hysteresis loops can be used to reveal the internal damage evolution in CMCs (Fantozzi and Reynaud 2009). Cho et al. (1991) developed an approach to estimate interface shear stress from frictional heating measurement. By analyzing the frictional heating data, Holmes and Cho [0] found that the interfacial shear stress of unidirectional SiC/CAS-II composite undergoes an initially rapid decrease at the initial stage of cyclic fatigue loading, i.e. from an initial value of over 20 MPa, to approximately 5 MPa after 25 000 cycles. Evans et al. (1995) developed an approach to evaluate interface shear stress by analyzing parabolic regions of hysteresis loops based on the Vagaggini's hysteresis loops models (Vagaggini et al. 1995). The initial interface shear stress of unidirectional SiC/CAS composite was approximately 20 MPa, and degraded to about 5 MPa at the 30th cycle. Li and Song (2010) and Li et al. (2013) developed an approach to estimate the interface shear stress of unidirectional CMCs. By comparing experimental hysteresis dissipated energy with theoretical values, the interface shear stress of unidirectional C/SiC composite has been estimated. The objective of this paper is to investigate the evolution of fiber/matrix interface shear stress of CMCs based on fatigue hysteresis loops, to reveal the internal fatigue damage evolution.

In this chapter, the hysteresis dissipated energy for the strain energy lost per volume during corresponding cycle is formulated in terms of interface shear stress. Comparing experimental fatigue hysteresis dissipated energy with theoretical computational values, the interface shear stress of unidirectional, cross-ply, 2D, and 3D CMCs at room temperature, 600, 800, 1000, 1200, and 1300 °C in inert, air, and steam conditions, are obtained. The effects of test temperature, oxidation, and fiber preforms on the degradation rate of interface shear stress are investigated, and the comparisons of interface degradation between C/SiC and SiC/SiC composites are analyzed.

4.2 Interface Damage Law at Room Temperature

Holmes and Cho (1992) investigated the tension–tension fatigue behavior of unidirectional SiC/CAS composite at room temperature. The fatigue loading was conducted under load control using a sinusoidal waveform, and a frequency of $f = 25$ Hz. The fatigue peak and valley stresses were $\sigma_{max} = 180$ MPa and $\sigma_{min} = 10$ MPa, respectively.

Under $\sigma_{max} = 180$ MPa, the fatigue hysteresis dissipated energy corresponding to different cycle number is shown in Figure 4.1a. The fatigue hysteresis dissipated energy corresponding to the 3rd, 10th, 20th, and 3200th applied cycles are 7.8, 9, 16, and 25 kPa, respectively. The interface shear stress corresponding to different

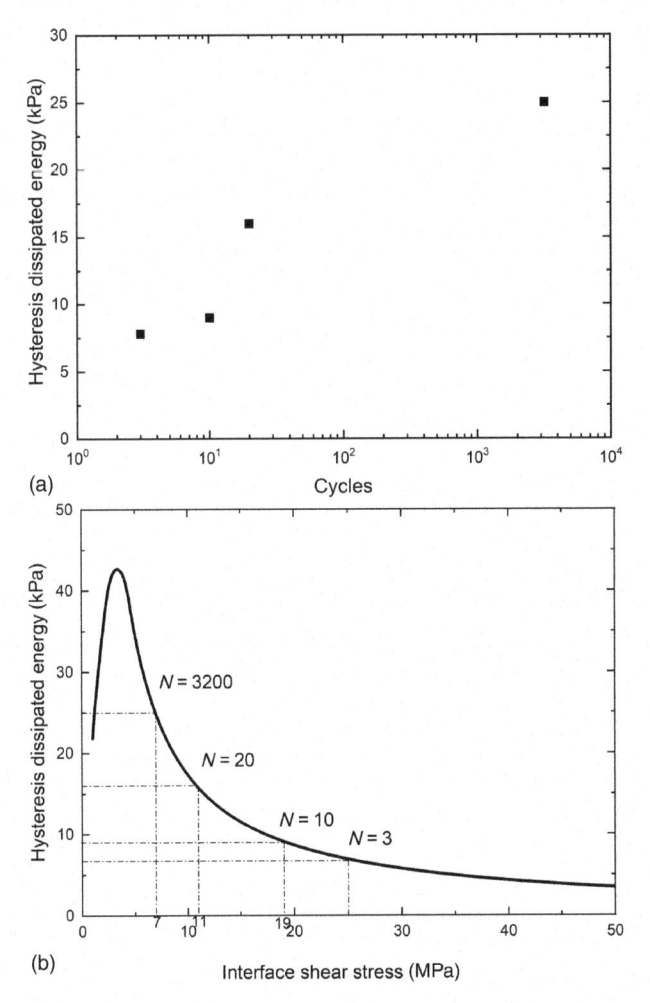

Figure 4.1 (a) The experimental fatigue hysteresis dissipated energy versus applied cycles; and (b) the theoretical fatigue hysteresis dissipated energy versus interface shear stress curve of unidirectional SiC/CAS composite under cyclic fatigue of $\sigma_{max} = 180$ MPa at room temperature.

4 Interface Damage Law of Ceramic-Matrix Composites

Table 4.1 The interface shear stress of unidirectional SiC/CAS composite under cyclic fatigue loading of $\sigma_{max} = 180$ MPa at room temperature.

Cycle number	Experimental hysteresis dissipated energy (kPa)	Interface shear stress (MPa)
3	7.8	25
10	9	19
20	16	11
3200	25	7

cycle number can be obtained from the hysteresis dissipated energy versus interface shear stress diagram, as shown in Figure 4.1b, by comparing the experimental fatigue hysteresis dissipated energy with theoretical computational values. The estimated interface shear stresses corresponding to different applied cycles are listed in Table 4.1.

The experimental and theoretical interface shear stress as a function of cycle number at room temperature is shown in Figure 4.2a. The interface shear stress decreases from 25 MPa at the 3rd cycle to 7 MPa at the 3200th cycle. The good matching with theoretical evolution in Figure 4.2a is linked to the shape of the law given in Eq. (4.1) (Evans et al. 1995).

$$\tau_i(N) = \tau_{io} + (1 - \exp(-\omega N^\lambda))(\tau_{imin} - \tau_{io}) \tag{4.1}$$

where τ_{io} denotes the initial interface shear stress, i.e. $\tau_i(N)$ at $N=1$, before fatigue loading; τ_{imin} denotes the steady-state interface shear stress during cycling; and ω and λ are empirical constants. The experimental and theoretical fatigue hysteresis dissipated energy versus cycle number curve is shown in Figure 4.2b. The fatigue hysteresis dissipated energy first increases with increasing cycle number from 7.8 kPa at the 1st cycle to the peak value of 24.5 kPa at the 48th cycle and then remains to be constant to 4000th cycle. The evolution of interface shear stress and fatigue hysteresis dissipated energy versus cycle number obtained using the present analysis agreed with experimental data, as shown in Figure 4.2a,b, respectively.

Opalski and Mall (1994) investigated the tension–tension fatigue behavior of cross-ply SiC/CAS composite at room temperature. The fatigue loading was in a sinusoidal waveform, and a frequency of $f = 10$ Hz. The fatigue peak and valley stresses were $\sigma_{max} = 180$ MPa and $\sigma_{min} = 18$ MPa, respectively.

Under the fatigue peak stress of $\sigma_{max} = 180$ MPa, the fatigue hysteresis dissipated energy corresponding to different cycle number is shown in Figure 4.3a. The fatigue hysteresis dissipated energy corresponding to the 10th, 100th, and 1000th applied cycles are 40, 48, and 58 kPa, respectively. The interface shear stress corresponding to different applied cycles can be obtained from the hysteresis dissipated energy versus interface shear stress diagram, as shown in Figure 4.3b, by comparing experimental fatigue hysteresis dissipated energy with theoretical computational values. The estimated interface shear stresses corresponding to different applied cycles are listed in Table 4.2.

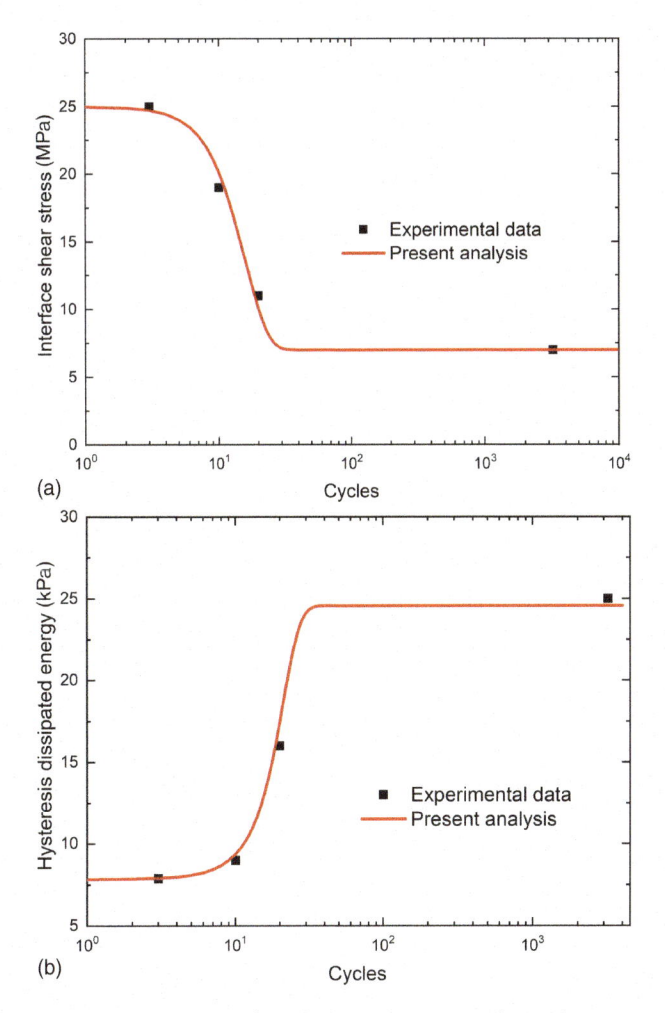

Figure 4.2 (a) The interface shear stress versus cycle number curve; and (b) the fatigue hysteresis dissipated energy versus cycle number curve of unidirectional SiC/CAS composite under cyclic fatigue of σ_{max} = 180 MPa at room temperature.

The experimental and theoretical interface shear stress as a function of cycle number at room temperature is shown in Figure 4.4a. The interface shear stress decreases from 18 MPa at the 10th cycle to 12 MPa at the 1000th cycle. The good matching with theoretical evolution in Figure 4.4a is linked to the shape of the law given in Eq. (4.1). The experimental and theoretical fatigue hysteresis dissipated energy versus cycle number curve is shown in Figure 4.4b. The fatigue hysteresis dissipated energy first increases with increasing cycle number from 39.5 kPa at the 1st cycle to the peak value of 54.6 kPa at the 568th cycle and then remains to be constant to 5000th cycle. The evolution of interface shear stress and fatigue hysteresis dissipated energy versus cycle number obtained using the present analysis agreed with experimental data, as shown in Figure 4.4a,b, respectively.

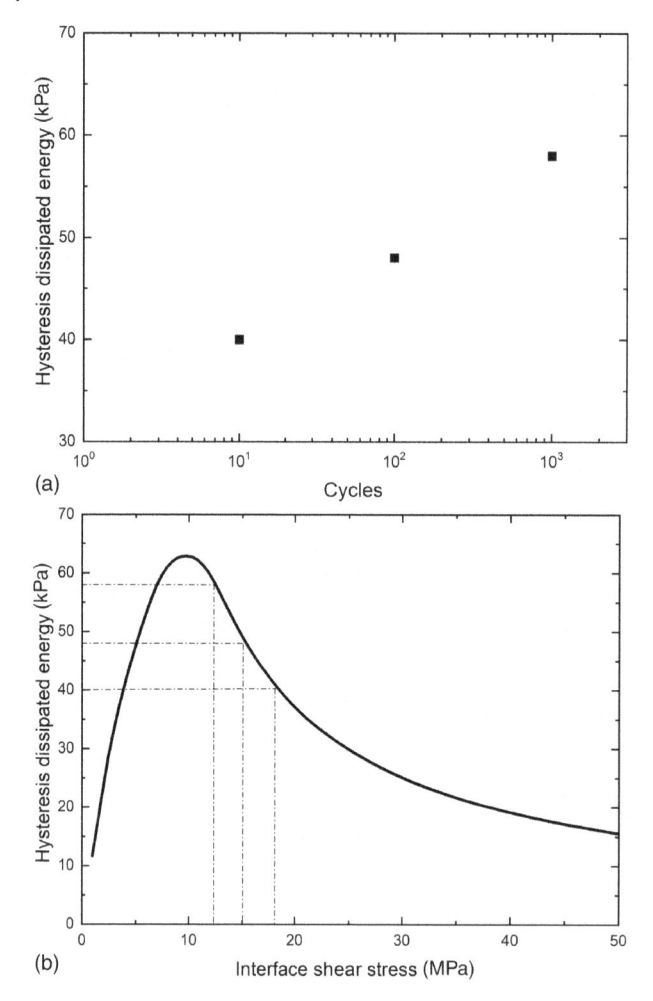

Figure 4.3 (a) The experimental fatigue hysteresis dissipated energy versus applied cycles; and (b) the theoretical fatigue hysteresis dissipated energy versus interface shear stress curve of cross-ply SiC/CAS composite under cyclic fatigue of σ_{max} = 180 MPa at room temperature.

Table 4.2 The interface shear stress of cross-ply SiC/CAS composite under cyclic fatigue loading of σ_{max} = 180 MPa at room temperature.

Cycle number	Experimental hysteresis dissipated energy (kPa)	Interface shear stress (MPa)
10	40	18
100	48	15
1000	58	12

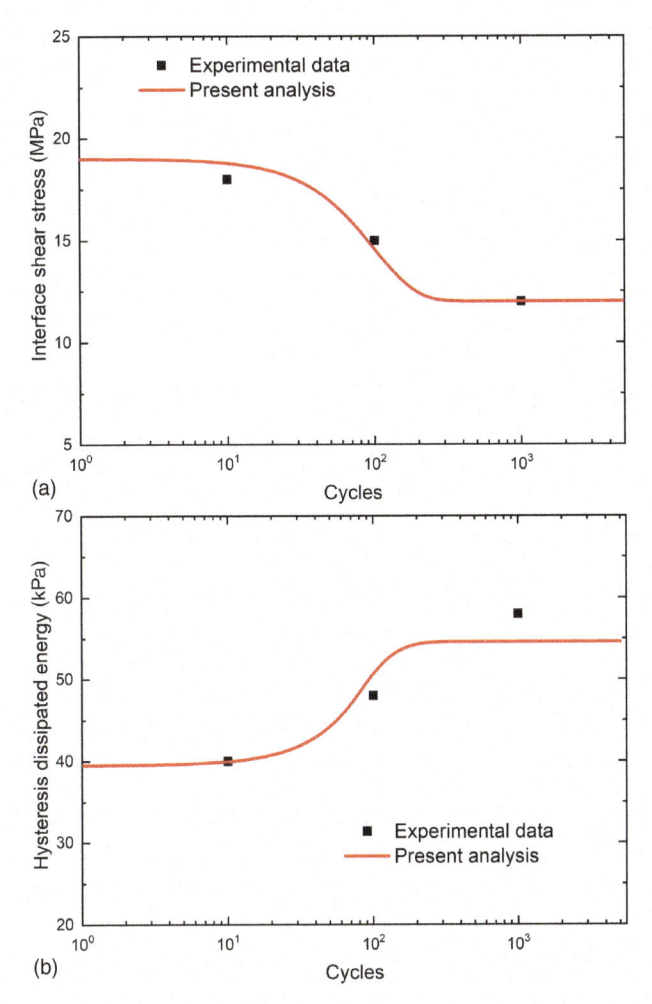

Figure 4.4 (a) The interface shear stress versus cycle number curve; and (b) the fatigue hysteresis dissipated energy versus cycle number curve of cross-ply SiC/CAS composite under cyclic fatigue of $\sigma_{max} = 180$ MPa at room temperature.

4.3 Interface Damage Law at Elevated Temperature in Inert Atmosphere

Reynaud (1996) investigated the tension–tension fatigue behavior of 2D SiC/SiC composite at 600, 800, and 1000 °C in inert atmosphere. The fatigue peak stress was $\sigma_{max} = 130$ MPa, and the valley stress was $\sigma_{min} = 0$ MPa. The loading frequency was $f = 1$ Hz.

The fatigue hysteresis dissipated energy corresponding to different applied cycles at 600, 800, and 1000 °C in inert atmosphere is shown in Figure 4.5a. The fatigue hysteresis dissipated energy increases with the cycle number and at the same cycle number increases with the test temperature. The interface

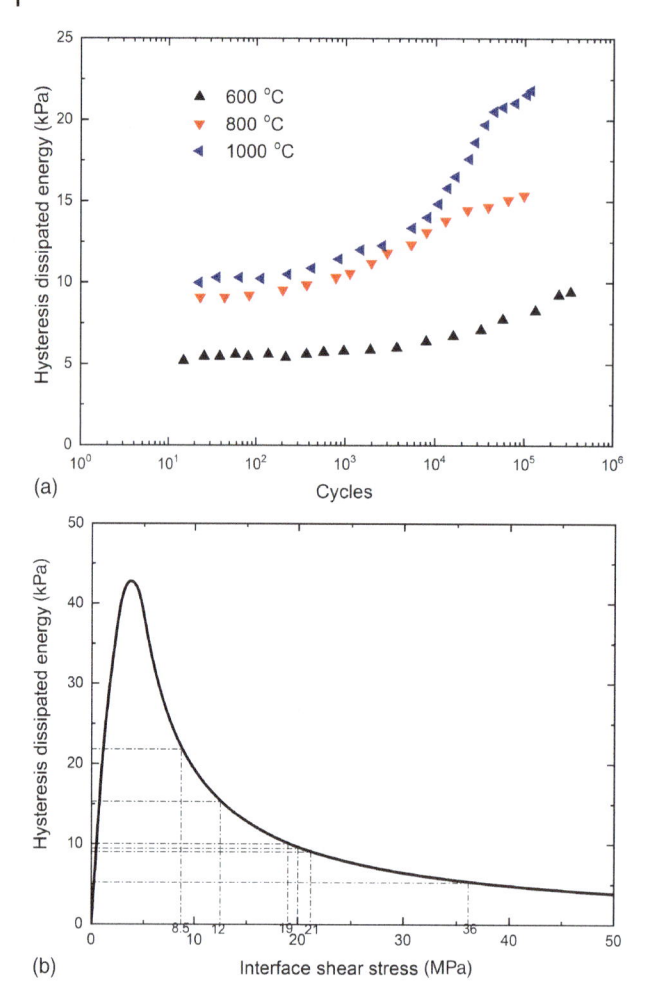

Figure 4.5 (a) The fatigue hysteresis dissipated energy versus cycle number curves corresponding to different experimental temperatures of 600, 800, and 1000 °C in inert atmosphere; and (b) the theoretical fatigue hysteresis dissipated energy versus cycle number curve of 2D SiC/SiC composite under cyclic fatigue of σ_{max} = 130 MPa at elevated temperature.

shear stress corresponding to different applied cycles can be obtained from the hysteresis dissipated energy versus interface shear stress diagram, as shown in Figure 4.5b, by comparing the experimental fatigue hysteresis dissipated energy with theoretical computational values. The estimated interface shear stresses corresponding to different applied cycles at 600, 800, and 1000 °C are listed in Tables 4.3–4.5, respectively.

The experimental and theoretical interface shear stress as a function of cycle number at 600 °C is shown in Figure 4.6a. The interface shear stress decreases from 35 MPa at the 1st cycle to 20.4 MPa at the 333 507th cycle. The good match-

Table 4.3 The interface shear stress of 2D SiC/SiC composite under cyclic fatigue loading of $\sigma_{max} = 130$ MPa at 600 °C in inert atmosphere.

Cycle number	Experimental hysteresis dissipated energy (kPa)	Interface shear stress (MPa)
25	5.4	35
58	5.6	34
577	5.7	33.6
1 917	5.8	32.8
3 821	6	32.1
8 014	6.3	30.1
15 970	6.7	28.7
32 648	7.1	27.1
57 260	7.7	24.8
132 997	8.2	23.3
245 481	9.2	20.9
333 507	9.4	20.4

Table 4.4 The interface shear stress of 2D SiC/SiC composite under cyclic fatigue loading of $\sigma_{max} = 130$ MPa at 800 °C in inert atmosphere.

Cycle number	Experimental hysteresis dissipated energy (kPa)	Interface shear stress (MPa)
23	9	21
82	9.2	20.9
202	9.5	20.2
374	9.8	19.5
784	10.2	18.7
1 121	10.5	18.2
1 967	11.2	17.2
2 960	11.8	16.3
5 464	12.3	15.6
8 014	13.1	14.7
13 019	13.8	13.9
22 834	14.4	13.3
39 038	14.6	13.1
65 059	15	12.8
97 894	15.3	12.5

Table 4.5 The interface shear stress of 2D SiC/SiC composite under cyclic fatigue loading of $\sigma_{max} = 130$ MPa at 1000 °C in inert atmosphere.

Cycle number	Experimental hysteresis dissipated energy (kPa)	Interface shear stress (MPa)
425	10.8	17.7
847	11.4	16.8
1 485	12	16
2 605	12.2	15.7
5 605	13.3	14.4
8 221	14	13.7
10 888	14.8	13
13 701	15.7	12.2
16 807	16.4	11.7
24 031	17.5	11
28 010	18.6	10.3
36 159	19.6	9.8
45 503	20.5	9.4
57 260	20.7	9.2
77 793	21	9.1
117 055	21.8	8.5

ing with theoretical evolution in Figure 4.6a is linked to the shape of the law given in Eq. (4.1). The experimental and theoretical fatigue hysteresis dissipated energy versus cycle number curve is shown in Figure 4.6b. The fatigue hysteresis dissipated energy increases with increasing cycle number from 5.5 kPa at the 1st cycle to 9.6 kPa at the 400 000th cycle. The evolution of interface shear stress and fatigue hysteresis dissipated energy versus cycle number obtained using the present analysis agreed with experimental data, as shown in Figure 4.6a,b, respectively.

The experimental and theoretical interface shear stress as a function of cycle number at 800 °C is shown in Figure 4.7a. The interface shear stress decreases from 22 MPa at the 1st cycle to 12.5 MPa at the 97 894th cycle. The good matching with theoretical evolution in Figure 4.7a is linked to the shape of the law given in Eq. (4.1). The experimental and theoretical fatigue hysteresis dissipated energy versus cycle number curve is shown in Figure 4.7b. The fatigue hysteresis dissipated energy first increases with the increase of cycle number from 9.2 kPa at the 1st cycle to the peak value of 15.4 kPa at the 122 364th cycle and then remains to be constant to 400 000th cycle. The evolution of interface shear stress and fatigue hysteresis dissipated energy versus cycle number obtained using the present analysis agreed with experimental data, as shown in Figure 4.7a,b, respectively.

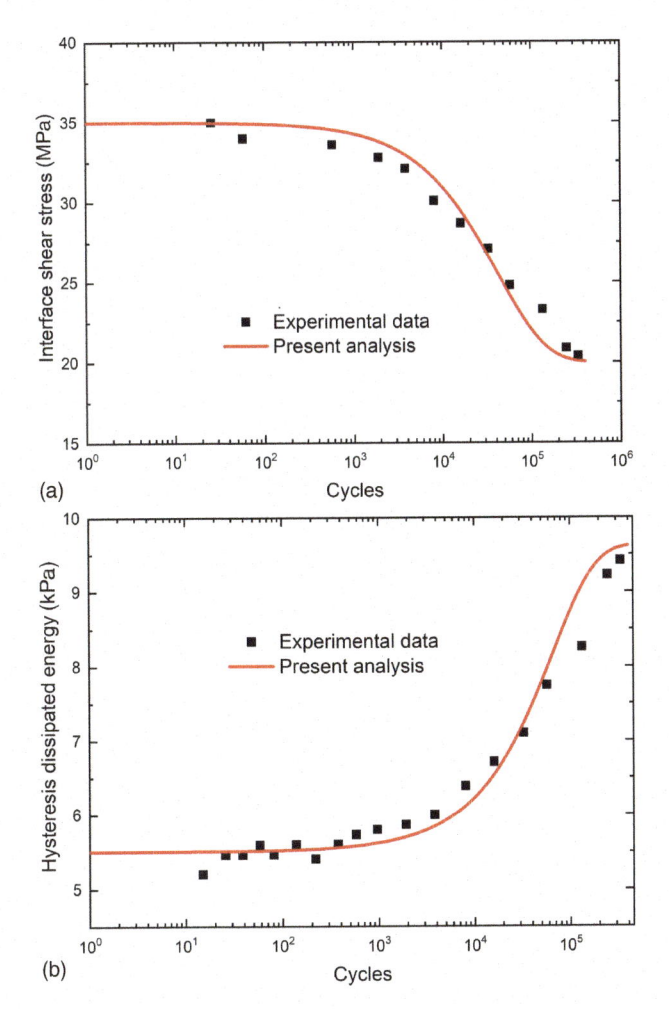

Figure 4.6 (a) The interface shear stress versus cycle number curve; and (b) the fatigue hysteresis dissipated energy versus cycle number curve of 2D SiC/SiC composite under cyclic fatigue of $\sigma_{max} = 130$ MPa at 600 °C in inert atmosphere.

The experimental and theoretical interface shear stress as a function of cycle number at 1000 °C is shown in Figure 4.8a. The interface shear stress decreases from 18 MPa at the 1st cycle to 8.5 MPa at the 117 055th cycle. The good matching with theoretical evolution in Figure 4.8a is linked to the shape of the law given in Eq. (4.1). The experimental and theoretical fatigue hysteresis dissipated energy versus cycle number curve is shown in Figure 4.8b. The fatigue hysteresis dissipated energy increases with increasing cycle number from 10.7 kPa at the 1st cycle to the peak value of 22.6 kPa at the 375 365th cycle and then remains to be constant to 400 000th cycle. The evolution of interface shear stress and fatigue hysteresis dissipated energy versus cycle number obtained using the present analysis agreed with experimental data, as shown in Figure 4.8a,b, respectively.

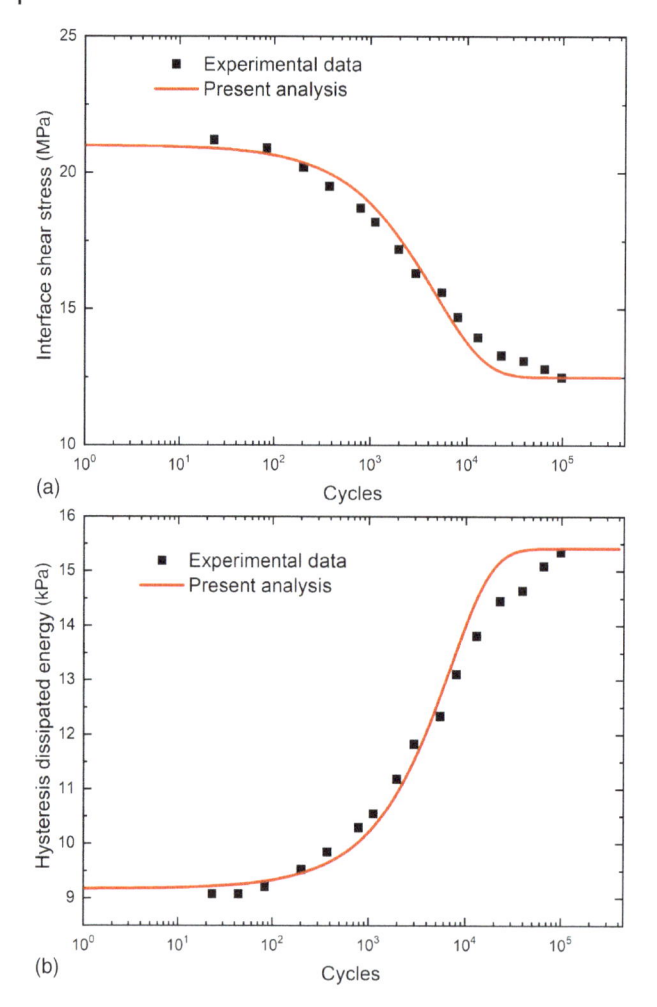

Figure 4.7 (a) The interface shear stress versus cycle number curve; and (b) the fatigue hysteresis dissipated energy U versus cycle number curve of 2D SiC/SiC composite under cyclic fatigue of $\sigma_{max} = 130$ MPa at 800 °C in inert atmosphere.

4.4 Interface Damage Law at Elevated Temperature in Air Atmosphere

4.4.1 1000 °C

Michael (2010) investigated the tension–tension fatigue behavior of 2D SiC/SiC composite at 1000 °C in air condition. The fatigue tests were conducted at the loading frequency of $f = 1$ Hz with a stress ratio of $R = 0.1$.

Under the fatigue peak stress of $\sigma_{max} = 80$ MPa at 1000 °C in air condition, the fatigue hysteresis dissipated energy corresponding to different cycle numbers is shown in Figure 4.9a. The fatigue hysteresis dissipated energy corresponding to

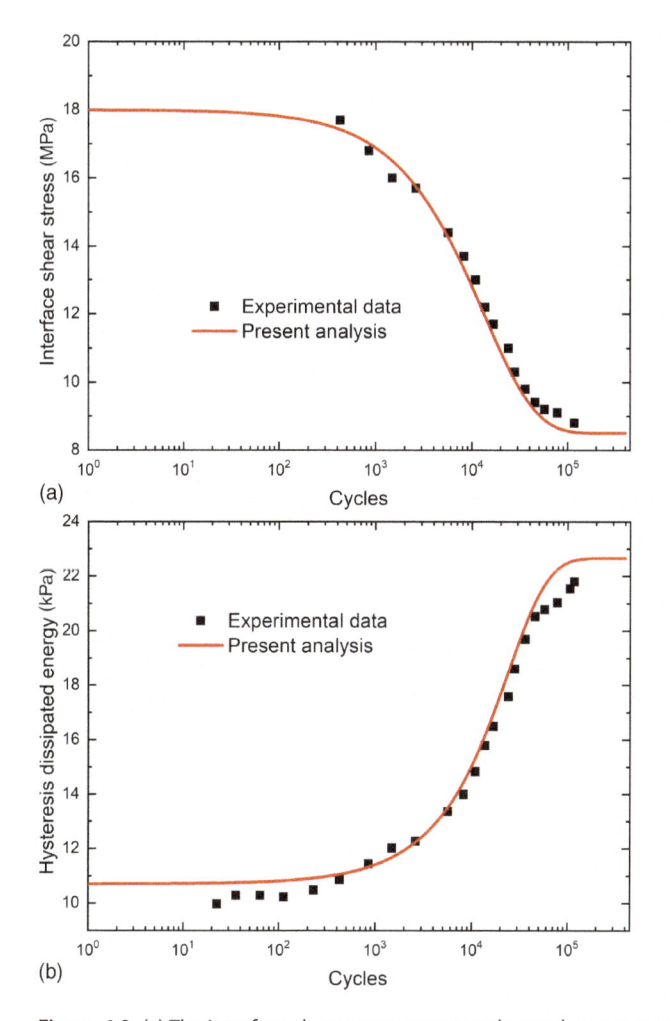

Figure 4.8 (a) The interface shear stress versus cycle number curve; and (b) the fatigue hysteresis dissipated energy versus cycle number curve of 2D SiC/SiC composite under cyclic fatigue of σ_{max} = 130 MPa at 1000 °C in inert atmosphere.

the 2nd, 1000th, 10 000th, 20 000th, and 30 000th applied cycles are 4.6, 5.4, 5.8, 6.3, and 7 kPa, respectively. The interface shear stress corresponding to different applied cycles can be obtained from the hysteresis dissipated energy versus interface shear stress diagram, as shown in Figure 4.9b, by comparing experimental fatigue hysteresis dissipated energy with theoretical computational values. The estimated interface shear stresses corresponding to different applied cycles are listed in Table 4.6.

The experimental and theoretical interface shear stress as a function of cycle number at 1000 °C in air condition is shown in Figure 4.10a. The interface shear stress decreases from 15 MPa at the 2nd cycle to 10 MPa at the 30 000th cycle. The good matching with theoretical evolution in Figure 4.10a is linked to the shape of the law given in Eq. (4.1). The experimental and theoretical fatigue

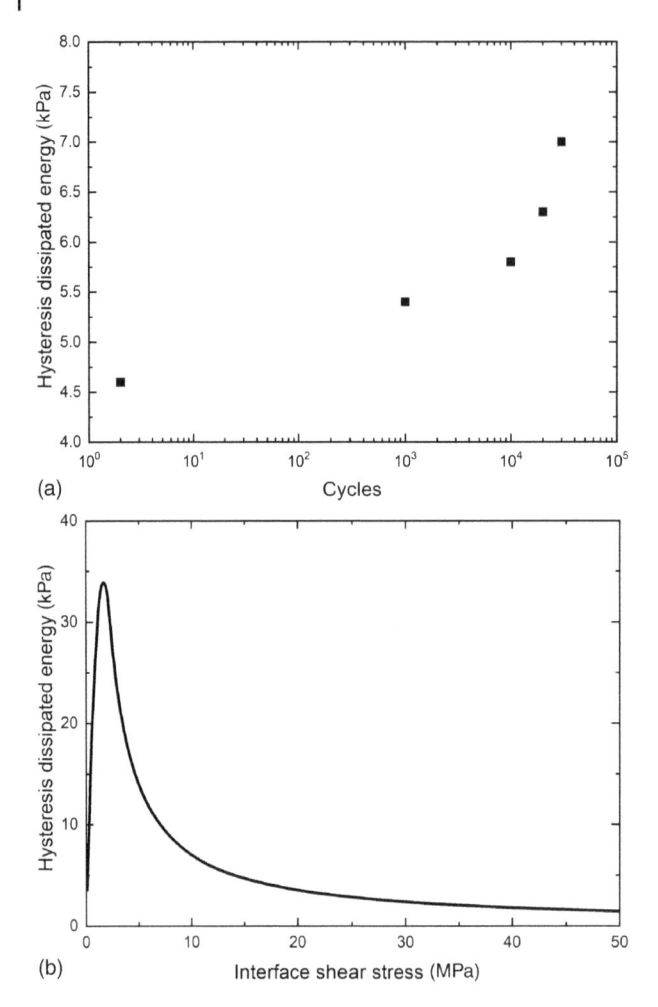

Figure 4.9 (a) The experimental fatigue hysteresis dissipated energy versus applied cycles; and (b) the theoretical fatigue hysteresis dissipated energy versus interface shear stress curve of 2D SiC/SiC composite under cyclic fatigue of $\sigma_{max} = 80$ MPa at 1000 °C in air condition.

Table 4.6 The interface shear stress of 2D SiC/SiC composite under cyclic fatigue loading of $\sigma_{max} = 80$ MPa at 1000 °C in air condition.

Cycle number	Experimental hysteresis dissipated energy (kPa)	Interface shear stress (MPa)
2	4.6	15
1 000	5.4	13
10 000	5.8	12
20 000	6.3	11
30 000	7	10

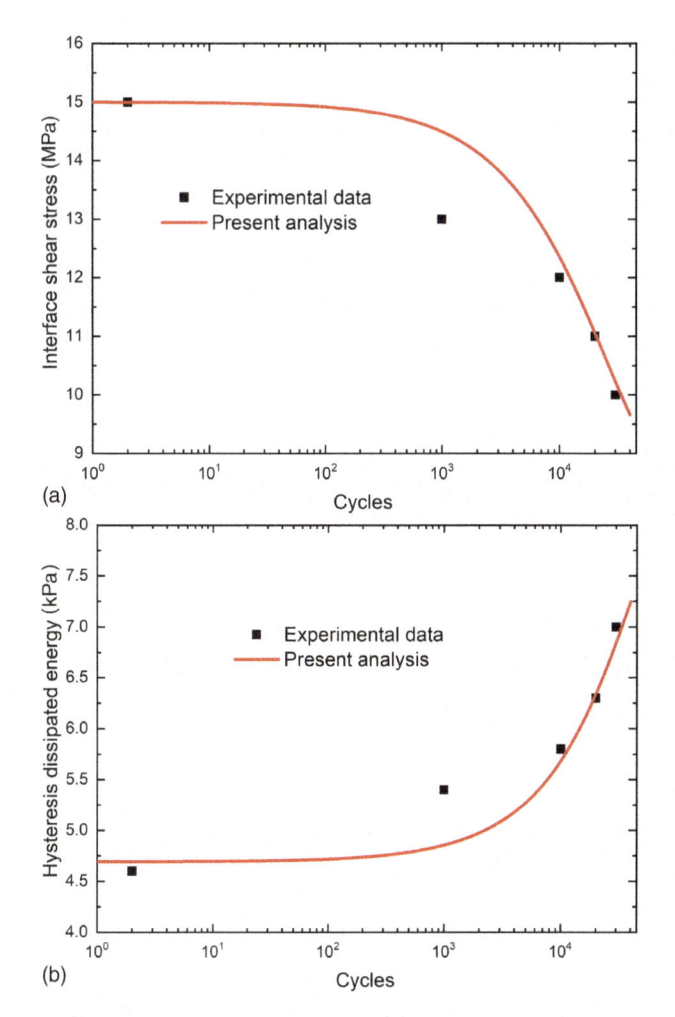

Figure 4.10 (a) The interface shear stress versus cycle number curve; and (b) the fatigue hysteresis dissipated energy versus cycle number curve of 2D SiC/SiC composite under cyclic fatigue of $\sigma_{max} = 80$ MPa at 1000 °C in air atmosphere.

hysteresis dissipated energy versus cycle number curve is shown in Figure 4.10b. The fatigue hysteresis dissipated energy increases with increasing cycle number from 4.7 kPa at the 1st cycle to 7.2 kPa at the 40 000th cycle. The evolution of interface shear stress and fatigue hysteresis dissipated energy versus cycle number obtained using the present analysis agreed with experimental data, as shown in Figure 4.10a,b, respectively.

4.4.2 1200 °C

Jacob (2010) investigated the tension–tension fatigue behavior of 2D SiC/SiC composite at 1200 °C in air condition. The fatigue tests were conducted at the loading frequency of $f = 0.1$ Hz with a stress ratio of $R = 0.05$.

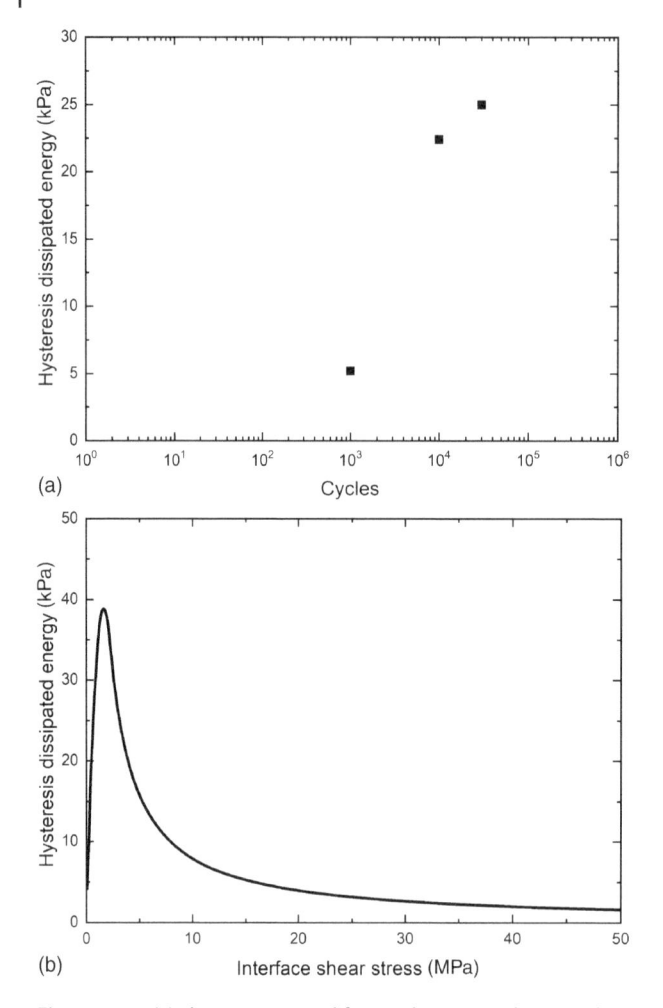

Figure 4.11 (a) The experimental fatigue hysteresis dissipated energy versus applied cycles; and (b) the theoretical fatigue hysteresis dissipated energy versus interface shear stress curve of 2D SiC/SiC composite under cyclic fatigue of σ_{max} = 140 MPa at 1200 °C in air condition.

Under the fatigue peak stress of σ_{max} = 140 MPa at 1200 °C in air condition, the fatigue hysteresis dissipated energy corresponding to different cycle numbers is shown in Figure 4.11a. The fatigue hysteresis dissipated energy corresponding to the 1000th, 10 000th, and 30 000th applied cycles are 5.2, 22.4, and 25 kPa, respectively. The interface shear stress corresponding to different applied cycles can be obtained from the hysteresis dissipated energy versus interface shear stress diagram, as shown in Figure 4.11b, by comparing experimental fatigue hysteresis dissipated energy with theoretical computational values. The estimated interface shear stresses corresponding to different applied cycles are listed in Table 4.7.

The experimental and theoretical interface shear stress as a function of cycle number at 1200 °C in air condition is shown in Figure 4.12a. The interface shear stress decreases from 15 MPa at the 1000th cycle to 3 MPa at the 30 000th cycle.

4.4 Interface Damage Law at Elevated Temperature in Air Atmosphere

Table 4.7 The interface shear stress of 2D SiC/SiC composite under cyclic fatigue loading of $\sigma_{max} = 60$ MPa at 1000 °C in steam.

Cycle number	Experimental hysteresis dissipated energy (kPa)	Interface shear stress (MPa)
2	1.5	15
10 000	2.3	10
100 000	2.9	8
150 000	4.6	5
190 000	7.7	3

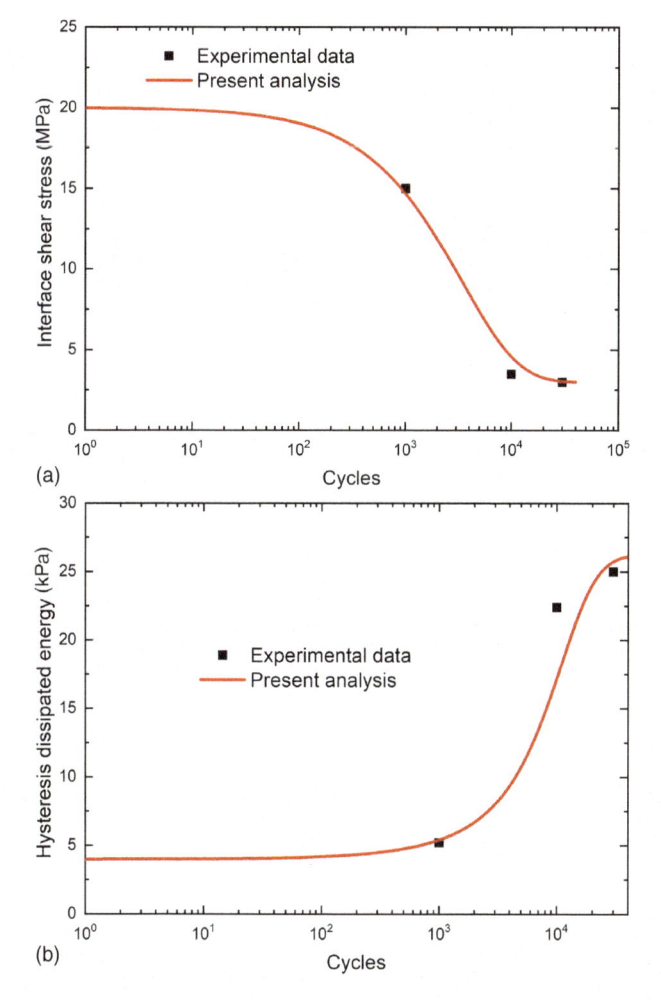

Figure 4.12 (a) The interface shear stress versus cycle number curve; and (b) the fatigue hysteresis dissipated energy versus cycle number curve of 2D SiC/SiC composite under cyclic fatigue of $\sigma_{max} = 140$ MPa at 1200 °C in air atmosphere.

The good matching with theoretical evolution in Figure 4.12a is linked to the shape of the law given in Eq. (4.1). The experimental and theoretical fatigue hysteresis dissipated energy versus cycle number curve is shown in Figure 4.12b. The fatigue hysteresis dissipated energy increases with increasing cycle number from 3.9 kPa at the 1st cycle to 26 kPa at the 40 000th cycle. The evolution of interface shear stress and fatigue hysteresis dissipated energy versus cycle number obtained using the present analysis agreed with experimental data, as shown in Figure 4.12a,b, respectively.

4.4.3 1300 °C

Zhu et al. (1998) investigated the tension–tension fatigue behavior of 2D SiC/SiC composite at 1300 °C in air condition. The fatigue tests were conducted with sinusoidal loading frequency of $f = 20$ Hz and stress ratio of $R = 0.1$.

Under the fatigue peak stress of $\sigma_{max} = 90$ MPa at 1300 °C in air condition, the fatigue hysteresis dissipated energy corresponding to different cycle numbers is shown in Figure 4.13a. The fatigue hysteresis dissipated energy corresponding to the 6000th, 24 000th, 90 000th, 650 000th, 1 200 000th, and 2 800 000th applied cycles are 2, 2.4, 2.9, 4.7, 5.9, and 7.8 kPa, respectively. The interface shear stress corresponding to different applied cycles can be obtained from the hysteresis dissipated energy versus interface shear stress diagram, as shown in Figure 4.13b, by comparing experimental fatigue hysteresis dissipated energy with theoretical computational values. The estimated interface shear stresses corresponding to different applied cycles are listed in Table 4.8.

The experimental and theoretical interface shear stress as a function of cycle number at 1300 °C in air condition is shown in Figure 4.14a. The interface shear stress decreases from 12 MPa at the 6000th cycle to 3 MPa at the 2 800 000th cycle. The good matching with theoretical evolution in Figure 4.14a is linked to the shape of the law given in Eq. (4.1). The experimental and theoretical fatigue hysteresis dissipated energy versus cycle number curve is shown in Figure 4.14b. The fatigue hysteresis dissipated energy increases with the increase of cycle number from 1.6 kPa at the 1st cycle to 7.9 kPa at the 3 000 000th cycle. The evolution of interface shear stress and fatigue hysteresis dissipated energy versus cycle number obtained using the present analysis agreed with experimental data, as shown in Figure 4.14a,b, respectively.

Under the fatigue peak stress of $\sigma_{max} = 120$ MPa at 1300 °C in air condition, the fatigue hysteresis dissipated energy corresponding to different applied cycle numbers is shown in Figure 4.15a. The fatigue hysteresis dissipated energy corresponding to the 100th, 6000th, 18 000th, and 36 000th applied cycles are 4, 7.2, 8.9, and 19 kPa, respectively. The interface shear stress corresponding to different applied cycles can be obtained from the hysteresis dissipated energy versus interface shear stress diagram, as shown in Figure 4.15b, by comparing experimental fatigue hysteresis dissipated energy with theoretical computational values. The estimated interface shear stresses corresponding to different applied cycles are listed in Table 4.9.

The experimental and theoretical interface shear stress as a function of cycle number at 1300 °C in air condition is shown in Figure 4.16a. The interface shear

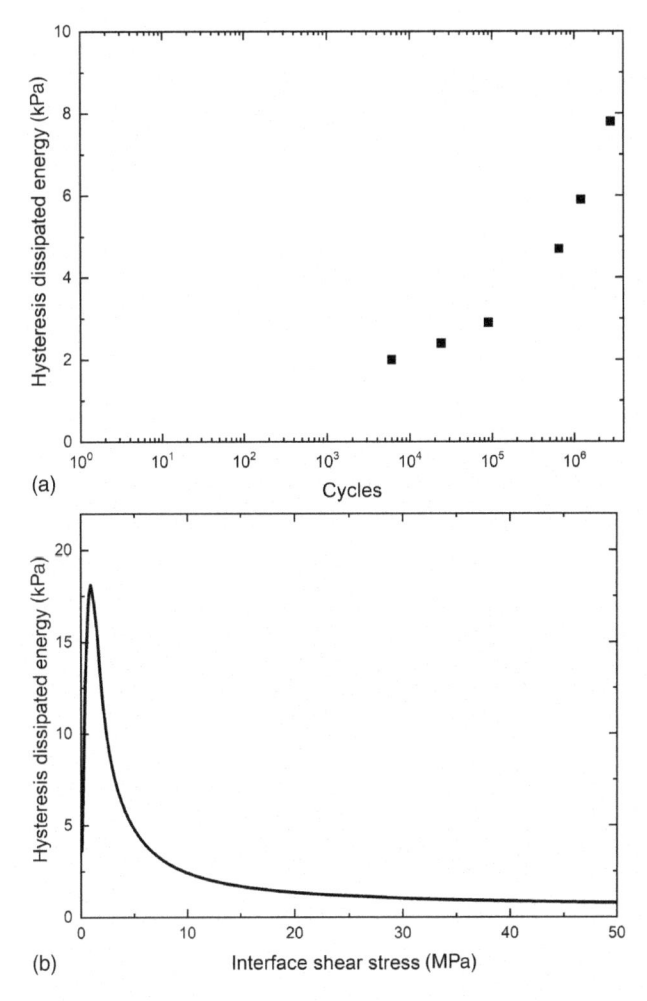

Figure 4.13 (a) The experimental fatigue hysteresis dissipated energy versus applied cycles; and (b) the theoretical fatigue hysteresis dissipated energy versus interface shear stress curve of 2D SiC/SiC composite under cyclic fatigue of σ_{max} = 90 MPa at 1300 °C in air condition.

Table 4.8 The interface shear stress of 2D SiC/SiC composite under cyclic fatigue loading of σ_{max} = 100 MPa at 1000 °C in steam condition.

Cycle number	Experimental hysteresis dissipated energy (kPa)	Interface shear stress (MPa)
2	9	15
500	10.4	13
3 000	13.5	10
10 000	16.8	8

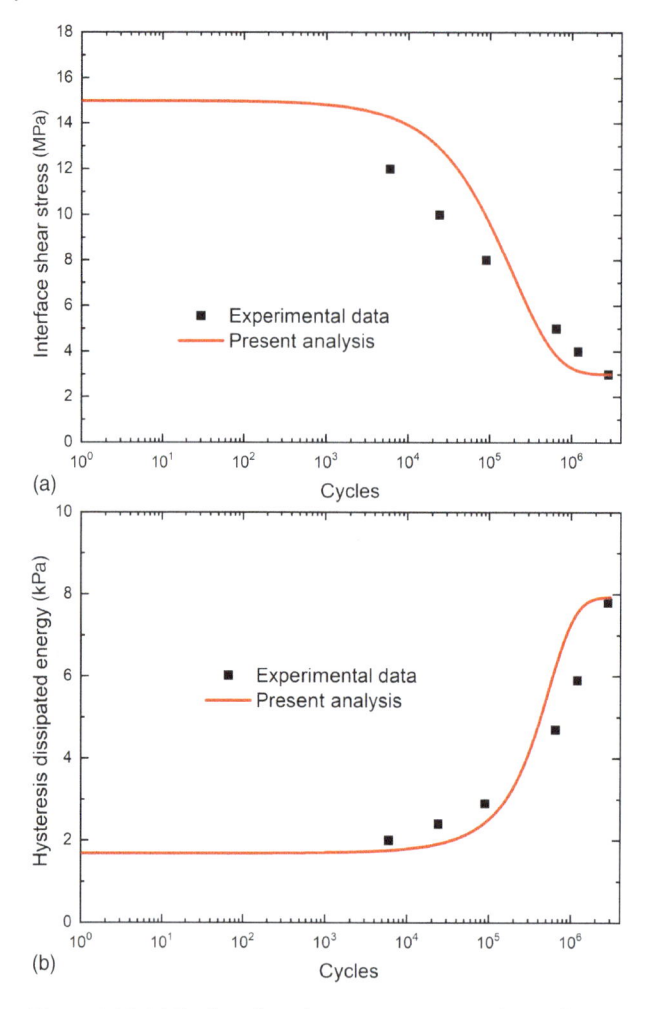

Figure 4.14 (a) The interface shear stress versus cycle number curve; and (b) the fatigue hysteresis dissipated energy versus cycle number curve of 2D SiC/SiC composite under cyclic fatigue of $\sigma_{max} = 90$ MPa at 1300 °C in air condition.

stress decreases from 18 MPa at the 100th cycle to 3.7 MPa at the 36 000th cycle. The good matching with theoretical evolution in Figure 4.16a is linked to the shape of the law given in Eq. (4.1). The experimental and theoretical fatigue hysteresis dissipated energy versus cycle number curve is shown in Figure 4.16b. The fatigue hysteresis dissipated energy increases with the increase of cycle number from 3.6 kPa at the 1st cycle to 15.4 kPa at the 40 000th cycle. The evolution of interface shear stress and fatigue hysteresis dissipated energy versus cycle number obtained using the present analysis agreed with experimental data, as shown in Figure 4.16a,b, respectively.

Shi et al. (2015) investigated the tension–tension fatigue behavior of 3D braided SiC/SiC composite at 1300 °C in air condition. The fatigue tests were conducted with sinusoidal loading frequency of $f = 1$ Hz and stress ratio of $R = 0.1$.

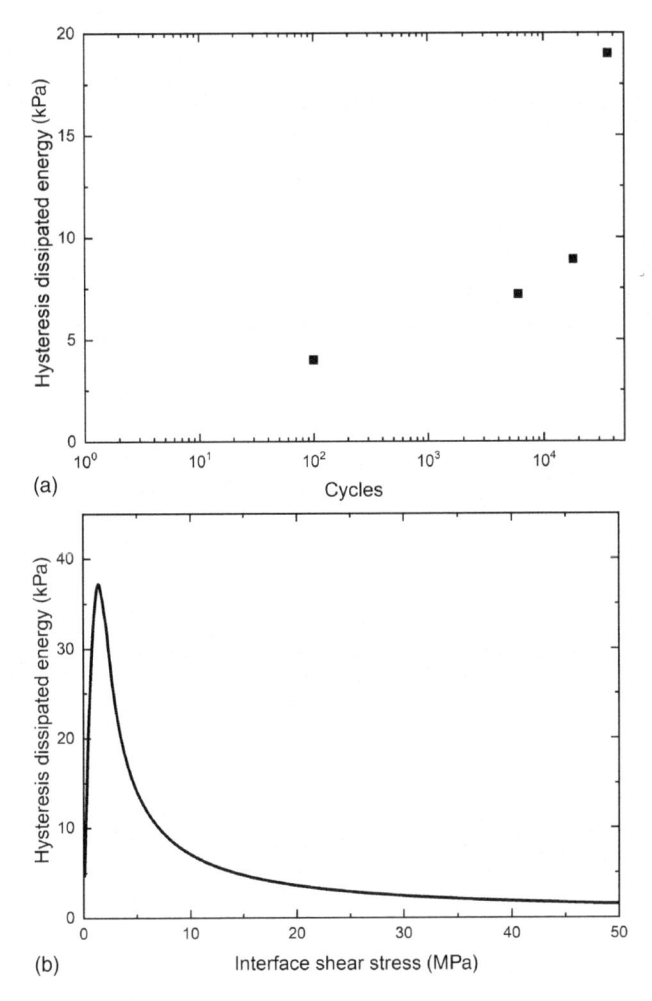

Figure 4.15 (a) The experimental fatigue hysteresis dissipated energy versus applied cycles; and (b) the theoretical fatigue hysteresis dissipated energy versus interface shear stress curve of 2D woven SiC/SiC composite under cyclic fatigue of σ_{max} = 120 MPa at 1300 °C in air condition.

Table 4.9 The interface shear stress of 2D SiC/SiC composite under cyclic fatigue loading of σ_{max} = 140 MPa at 1200 °C in air condition.

Cycle number	Experimental hysteresis dissipated energy (kPa)	Interface shear stress (MPa)
1 000	5.2	15
10 000	22.4	3.5
30 000	25	3

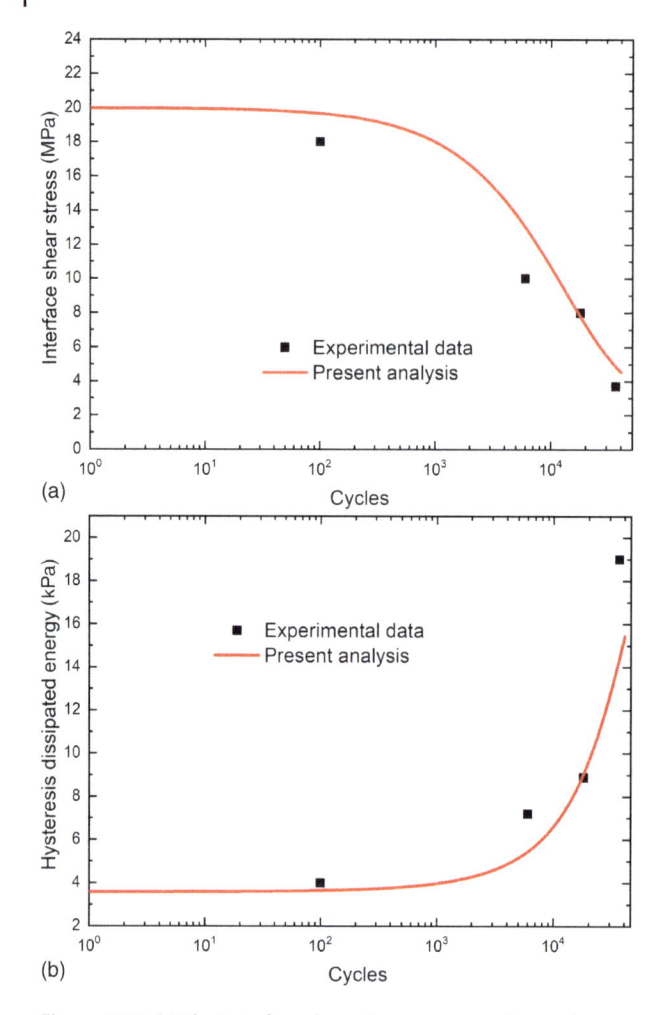

Figure 4.16 (a) The interface shear stress versus cycle number curve; and (b) the fatigue hysteresis dissipated energy versus cycle number curve of 2D SiC/SiC composite under cyclic fatigue of $\sigma_{max} = 120$ MPa at 1300 °C in air condition.

Under the fatigue peak stress of $\sigma_{max} = 100$ MPa at 1300 °C in air condition, the fatigue hysteresis dissipated energy corresponding to different cycle numbers is shown in Figure 4.17a. The fatigue hysteresis dissipated energy corresponding to the 10th, 50th, 100th, 200th, 300th, and 400th applied cycles are 7.4, 8.3, 9.9, 12, 14.2, and 26 kPa, respectively. The interface shear stress corresponding to different applied cycles can be obtained from the hysteresis dissipated energy versus interface shear stress diagram, as shown in Figure 4.17b, by comparing experimental fatigue hysteresis dissipated energy with theoretical computational values. The estimated interface shear stresses corresponding to different applied cycles are listed in Table 4.10.

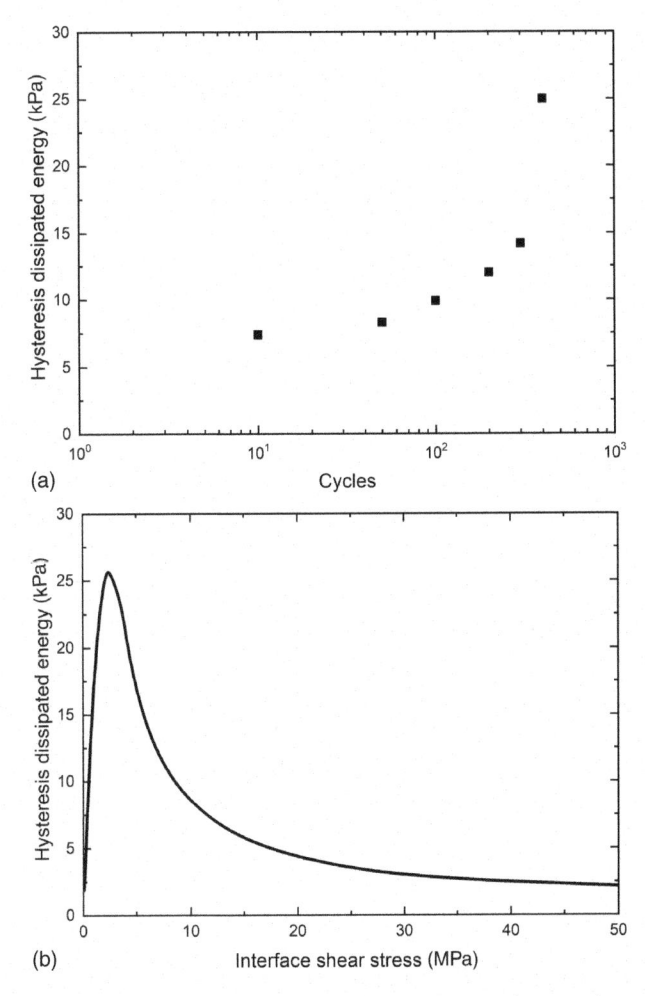

Figure 4.17 (a) The experimental fatigue hysteresis dissipated energy versus applied cycles; and (b) the theoretical fatigue hysteresis dissipated energy versus interface shear stress curve of 3D braided SiC/SiC composite under cyclic fatigue of σ_{max} = 100 MPa at 1300 °C in air condition.

Table 4.10 The interface shear stress of 2D SiC/SiC composite under cyclic fatigue loading of σ_{max} = 140 MPa at 1200 °C in steam condition.

Cycle number	Experimental hysteresis dissipated energy (kPa)	Interface shear stress (MPa)
100	4.5	17
1 000	19.3	4
10 000	24.6	3.2

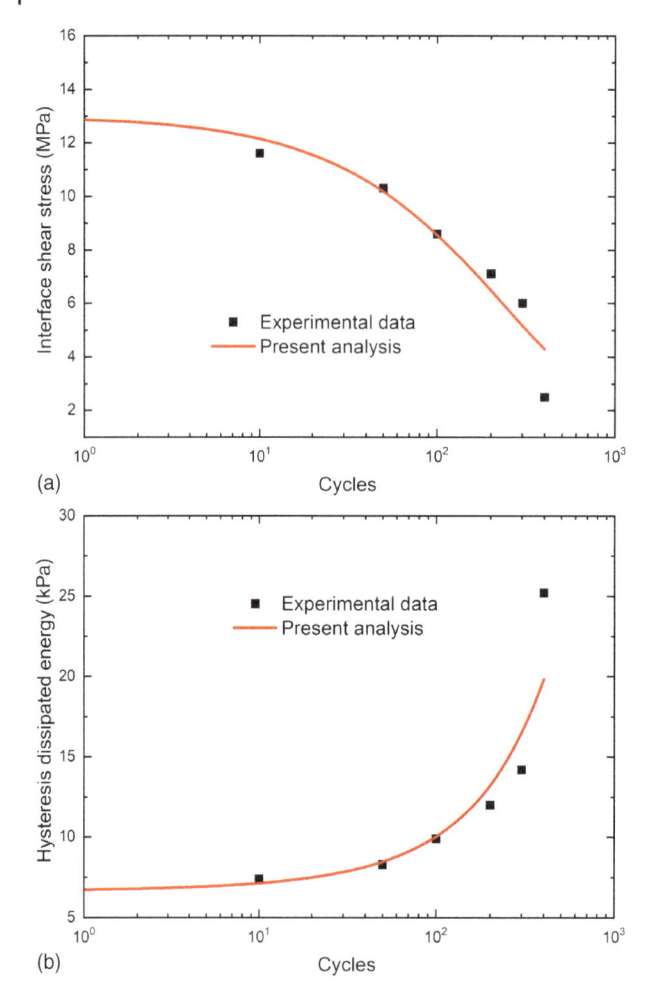

Figure 4.18 (a) The interface shear stress versus cycle number curve; and (b) the fatigue hysteresis dissipated energy versus cycle number curve of 3D braided SiC/SiC composite under cyclic fatigue of $\sigma_{max} = 100$ MPa at 1300 °C in air condition.

The experimental and theoretical interface shear stress as a function of cycle number at 1300 °C in air condition is shown in Figure 4.18a. The interface shear stress decreases from 11.6 MPa at the 10th cycle to 2.5 MPa at the 400th cycle. The good matching with theoretical evolution in Figure 4.18a is linked to the shape of the law given in Eq. (4.1). The experimental and theoretical fatigue hysteresis dissipated energy versus cycle number curve is shown in Figure 4.18b. The fatigue hysteresis dissipated energy increases with the increase of cycle number from 6.7 kPa at the 1st cycle to 20 kPa at the 400th cycle. The evolution of interface shear stress and fatigue hysteresis dissipated energy versus cycle number obtained using the present analysis agreed with experimental data, as shown in Figure 4.18a,b, respectively.

4.5 Interface Damage Law at Elevated Temperature in Steam Atmosphere

4.5.1 1000 °C

Michael (2010) investigated the tension–tension fatigue behavior of 2D SiC/SiC composite at 1000 °C in steam condition. The fatigue tests were conducted at the loading frequency of $f = 1$ Hz with a stress ratio of $R = 0.1$.

Under the fatigue peak stress of $\sigma_{max} = 60$ MPa at 1000 °C in steam condition, the fatigue hysteresis dissipated energy corresponding to different cycle numbers is shown in Figure 4.19a. The fatigue hysteresis dissipated energy corresponding

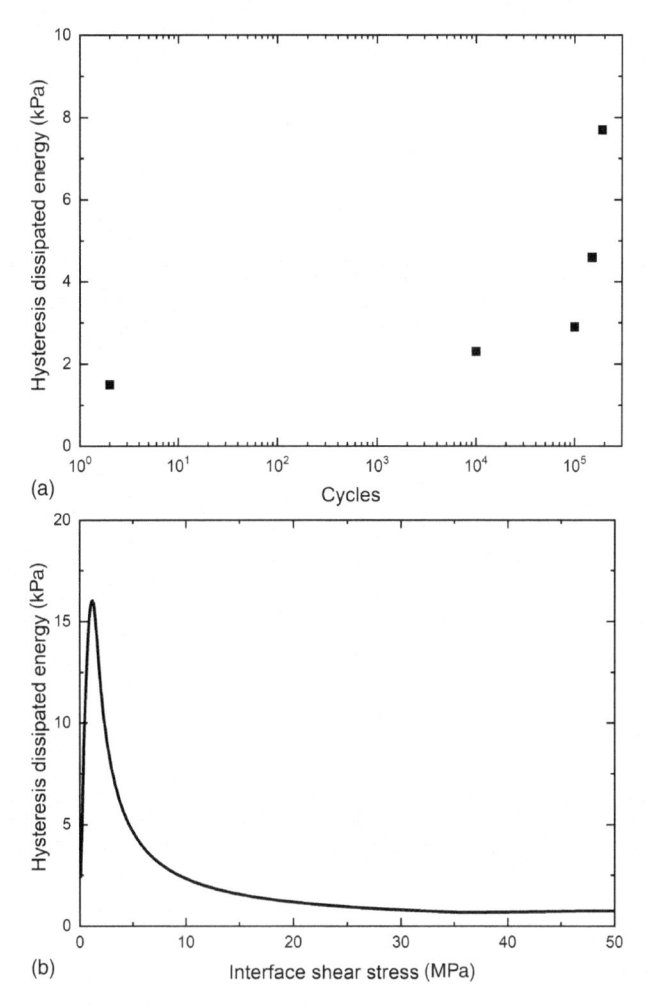

Figure 4.19 (a) The experimental fatigue hysteresis dissipated energy versus applied cycles; and (b) the theoretical fatigue hysteresis dissipated energy versus interface shear stress curve of 2D woven SiC/SiC composite under cyclic fatigue of $\sigma_{max} = 60$ MPa at 1000 °C in steam condition.

Table 4.11 The interface shear stress of 2D SiC/SiC composite under cyclic fatigue loading of $\sigma_{max} = 90$ MPa at 1300 °C in air condition.

Cycle number	Experimental hysteresis dissipated energy (kPa)	Interface shear stress (MPa)
6 000	2	12
24 000	2.4	10
90 000	2.9	8
650 000	4.7	5
1 200 000	5.9	4
2 800 000	7.8	3

to the 2nd, 10 000th, 100 000th, 150 000th, and 190 000th applied cycles are 1.5, 2.3, 2.9, 4.6, and 7.7 kPa, respectively. The interface shear stress corresponding to different applied cycles can be obtained from the hysteresis dissipated energy versus interface shear stress diagram, as shown in Figure 4.19b, by comparing experimental fatigue hysteresis dissipated energy with theoretical computational values. The estimated interface shear stresses corresponding to different applied cycles are listed in Table 4.11.

The experimental and theoretical interface shear stress as a function of cycle number at 1000 °C in steam condition is shown in Figure 4.20a. The interface shear stress decreases from 15 MPa at the 2nd cycle to 3 MPa at the 190 000th cycle. The good matching with theoretical evolution in Figure 4.20a is linked to the shape of the law given in Eq. (4.1). The experimental and theoretical fatigue hysteresis dissipated energy versus cycle number curve is shown in Figure 4.20b. The fatigue hysteresis dissipated energy increases with increasing cycle number from 1.5 kPa at the 1st cycle to 5.9 kPa at the 20 000th cycle. The evolution of interface shear stress and fatigue hysteresis dissipated energy versus cycle number obtained using the present analysis agreed with experimental data, as shown in Figure 4.20a,b, respectively.

Under the fatigue peak stress of $\sigma_{max} = 100$ MPa at 1000 °C in steam condition, the fatigue hysteresis dissipated energy corresponding to different cycle numbers is shown in Figure 4.21a. The fatigue hysteresis dissipated energy corresponding to the 2nd, 10 000th, 100 000th, 150 000th, and 190 000th applied cycles are 1.5, 2.3, 2.9, 4.6, and 7.7 kPa, respectively. The interface shear stress corresponding to different applied cycles can be obtained from the hysteresis dissipated energy versus interface shear stress diagram, as shown in Figure 4.21b, by comparing experimental fatigue hysteresis dissipated energy with theoretical computational values. The estimated interface shear stresses corresponding to different applied cycles are listed in Table 4.12.

The experimental and theoretical interface shear stress as a function of cycle number at 1000 °C in steam condition is shown in Figure 4.22a. The interface shear stress decreases from 15 MPa at the 2nd cycle to 8 MPa at the 10 000th cycle. The good matching with theoretical evolution in Figure 4.22a is linked to the shape of the law given in Eq. (4.1). The experimental and theoretical fatigue

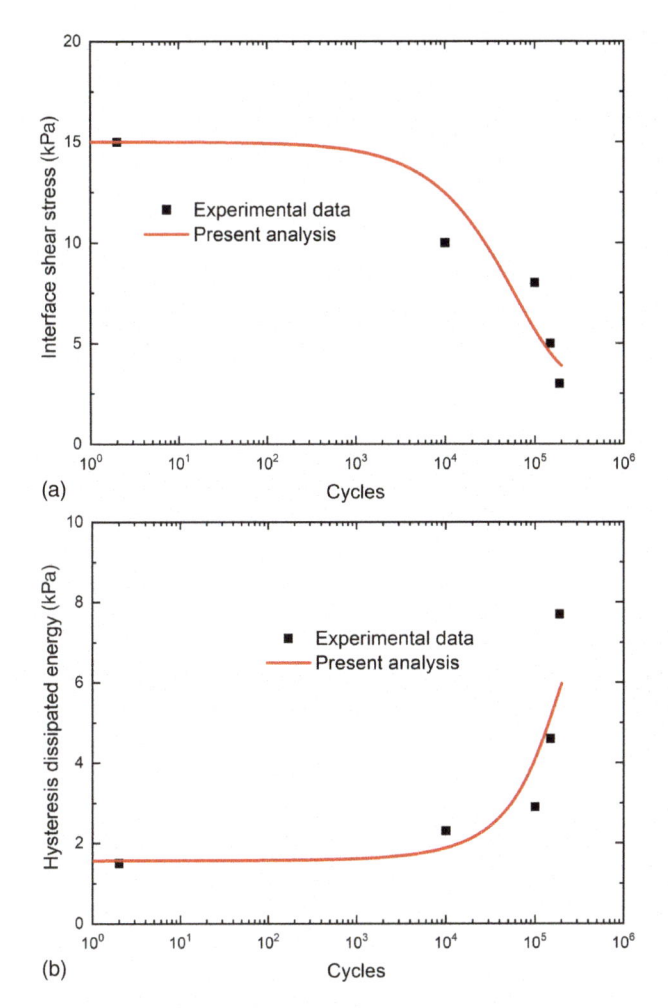

Figure 4.20 (a) The interface shear stress versus cycle number curve; and (b) the fatigue hysteresis dissipated energy versus cycle number curve of 2D SiC/SiC composite under cyclic fatigue of $\sigma_{max} = 60$ MPa at 1000 °C in steam condition.

hysteresis dissipated energy versus cycle number curve is shown in Figure 4.22b. The fatigue hysteresis dissipated energy increases with increasing cycle number from 9.1 kPa at the 1st cycle to 16.8 kPa at the 20 000th cycle. The evolution of interface shear stress and fatigue hysteresis dissipated energy versus cycle number obtained using the present analysis agreed with experimental data, as shown in Figure 4.22a,b, respectively.

4.5.2 1200 °C

Jacob (2010) investigated the tension–tension fatigue behavior of 2D SiC/SiC composite at 1200 °C in steam condition. The fatigue tests were conducted at the loading frequency of $f = 0.1$ Hz with a stress ratio of $R = 0.05$.

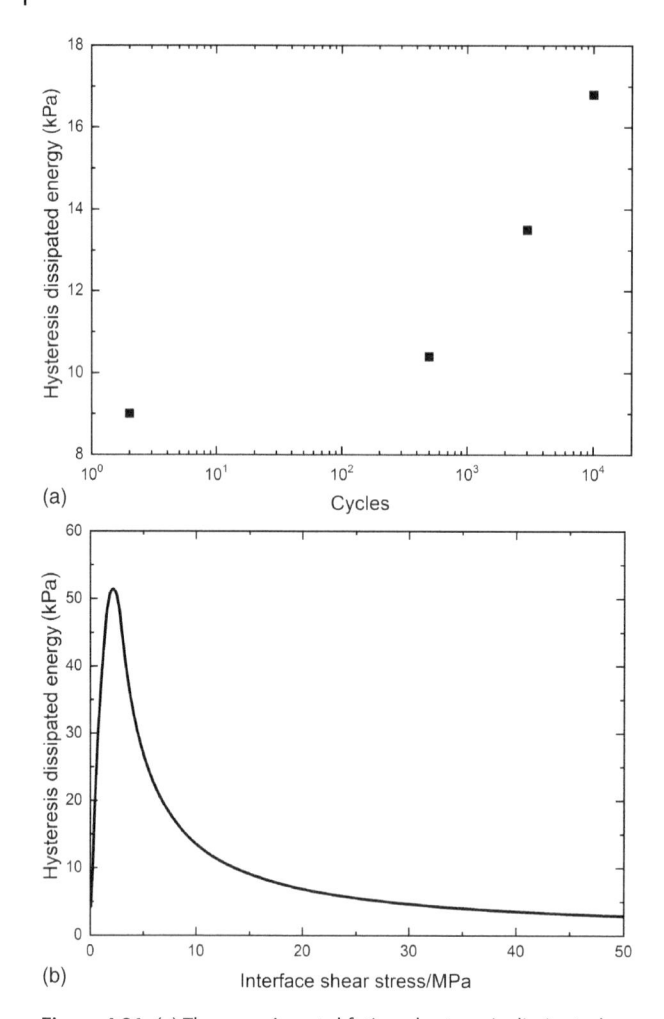

188 | *4 Interface Damage Law of Ceramic-Matrix Composites*

Figure 4.21 (a) The experimental fatigue hysteresis dissipated energy versus applied cycles; and (b) the theoretical fatigue hysteresis dissipated energy versus interface shear stress curve of 2D woven SiC/SiC composite under cyclic fatigue of $\sigma_{max} = 100$ MPa at 1000 °C in steam condition.

Table 4.12 The interface shear stress of 2D SiC/SiC composite under cyclic fatigue loading of $\sigma_{max} = 120$ MPa at 1300 °C in air condition.

Cycle number	Experimental hysteresis dissipated energy (kPa)	Interface shear stress (MPa)
100	4	18
6 000	7.2	10
18 000	8.9	8
36 000	19	3.7

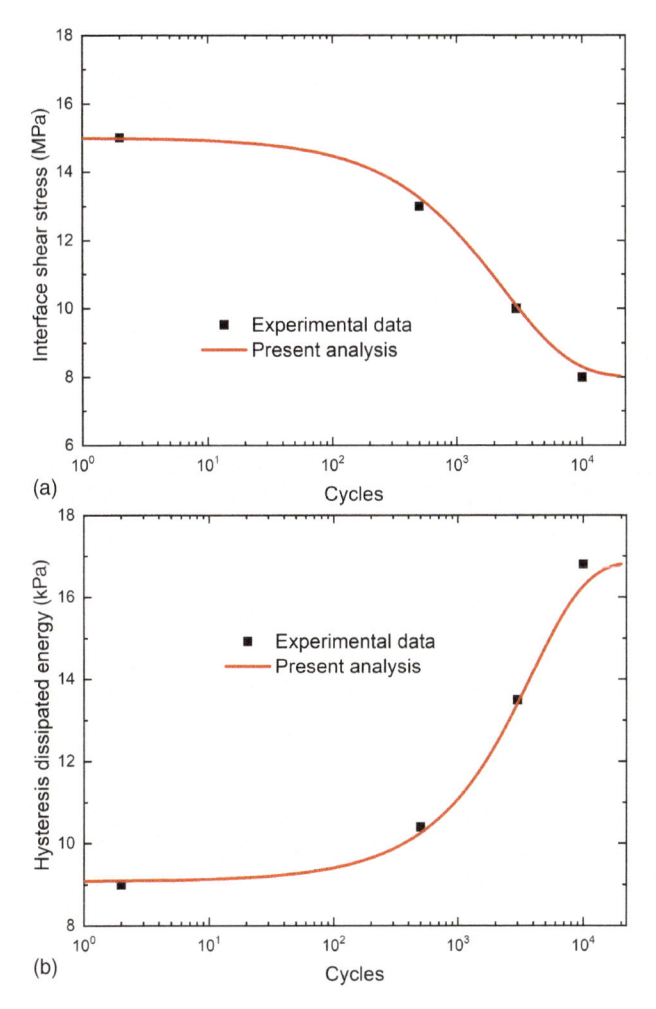

Figure 4.22 (a) The interface shear stress versus cycle number curve; and (b) the fatigue hysteresis dissipated energy versus cycle number curve of 2D SiC/SiC composite under cyclic fatigue of σ_{max} = 100 MPa at 1000 °C in steam condition.

Under the fatigue peak stress of σ_{max} = 140 MPa at 1200 °C in steam condition, the fatigue hysteresis dissipated energy corresponding to different cycle numbers is shown in Figure 4.23a. The fatigue hysteresis dissipated energy corresponding to the 100th, 1000th, and 10 000th applied cycles are 4.5, 19.3, and 24.6 kPa, respectively. The interface shear stress corresponding to different applied cycles can be obtained from the hysteresis dissipated energy versus interface shear stress diagram, as shown in Figure 4.23b, by comparing experimental fatigue hysteresis dissipated energy with theoretical computational values. The estimated interface shear stresses corresponding to different applied cycles are listed in Table 4.13.

The experimental and theoretical interface shear stress as a function of cycle number at 1200 °C in steam condition is shown in Figure 4.24a. The interface

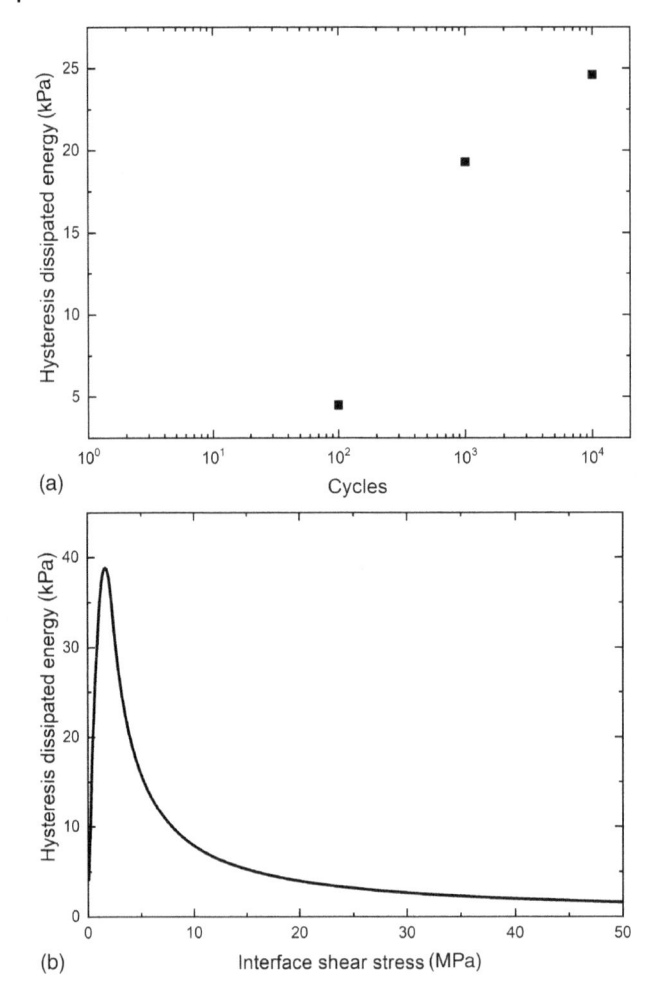

Figure 4.23 (a) The experimental fatigue hysteresis dissipated energy versus applied cycles; and (b) the theoretical fatigue hysteresis dissipated energy versus interface shear stress curve of 2D woven SiC/SiC composite under cyclic fatigue of $\sigma_{max} = 140$ MPa at 1200 °C in steam condition.

Table 4.13 The interface shear stress of 3D SiC/SiC composite under cyclic fatigue loading of $\sigma_{max} = 100$ MPa at 1300 °C in air condition.

Cycle number	Experimental hysteresis dissipated energy (kPa)	Interface shear stress (MPa)
10	7.4	11.6
50	8.3	10.3
100	9.9	8.6
200	12	7.1
300	14.2	6
400	25	2.5

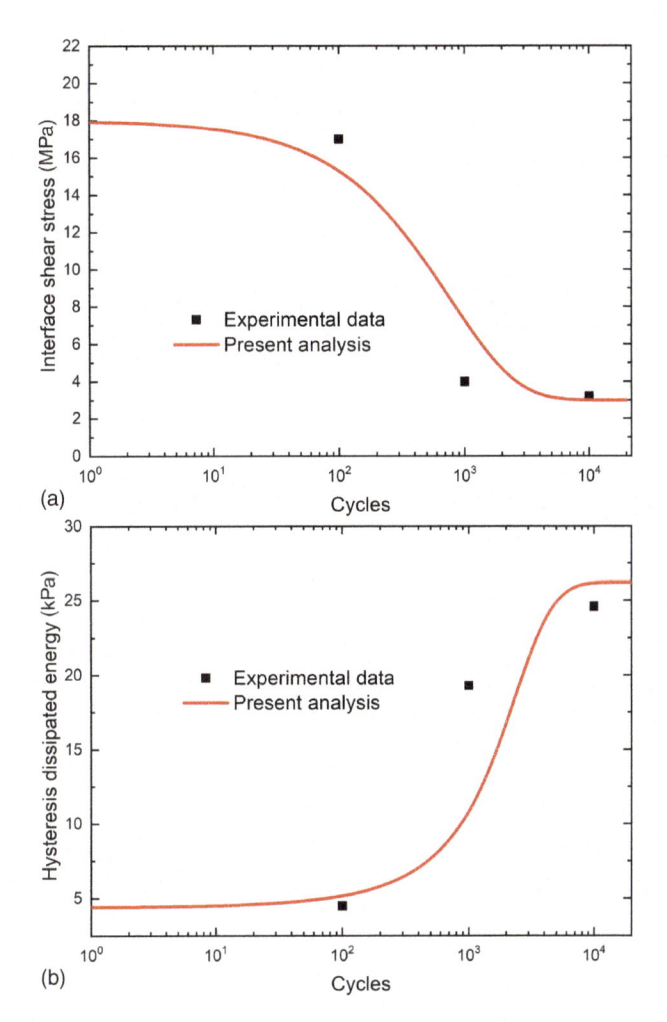

Figure 4.24 (a) The interface shear stress versus cycle number curve; and (b) the fatigue hysteresis dissipated energy versus cycle number curve of 2D SiC/SiC composite under cyclic fatigue of $\sigma_{max} = 140$ MPa at 1200 °C in steam condition.

shear stress decreases from 17 MPa at the 100th cycle to 3.2 MPa at the 10 000th cycle. The good matching with theoretical evolution in Figure 4.24a is linked to the shape of the law given in Eq. (4.1). The experimental and theoretical fatigue hysteresis dissipated energy versus cycle number curve is illustrated in Figure 4.24b. The fatigue hysteresis dissipated energy first increases with the increase of cycle number from 4.4 kPa at the 1st cycle to the peak value of 26.2 kPa at the 19 875th cycle and then remains to be constant to 20 000th cycle. The evolution of interface shear stress and fatigue hysteresis dissipated energy versus cycle number obtained using the present analysis agreed with experimental data, as shown in Figure 4.24a,b, respectively.

4.6 Results and Discussion

4.6.1 Effect of Temperature, Oxidation, and Fiber Preforms on Interface Damage of CMCs

The interface shear stress versus cycle number curves of unidirectional, cross-ply SiC/CAS, 2D, and 3D SiC/SiC composites at room and elevated temperatures are shown in Figure 4.25. The interface shear stress decreases with increasing cycle number, and the degradation rate is affected by test temperatures, oxidation, and fiber preforms.

For unidirectional and cross-ply SiC/CAS composite at room temperature, the interface shear stress degradation rate is higher corresponding to unidirectional SiC/CAS composite, i.e. the interface shear stress decreases 72% from the 3rd cycle to 3200th cycle, and lower for cross-ply SiC/CAS composite, i.e. the interface shear stress decreases 33.3% from the 10th cycle to the 1000th cycle.

For 2D SiC/SiC composite at 600, 800, and 1000 °C in inert atmosphere, the interface shear stress decreases with increasing test temperature, i.e. $\tau_i = 35$ MPa when $N = 25$ at 600 °C, and $\tau_i = 21$ MPa when $N = 23$ at 800 °C; and the interface shear stress degradation rate increases with increasing test temperature, i.e. the interface shear stress decreases 41.7% from 25th cycle to 333 507th cycle at 600 °C, and 40.4% from 23rd cycle to 97 894th cycle.

For 2D SiC/SiC composite at 1000 °C in inert, air, and steam conditions, the interface shear stress degradation rate is the highest corresponding to the steam condition, i.e. the interface shear stress decreases 46.6% from the 2nd cycle to 10 000th cycle, and the lowest for the inert condition, i.e. the interface shear stress decreases 51.9% from the 425th cycle to 117 055th cycle.

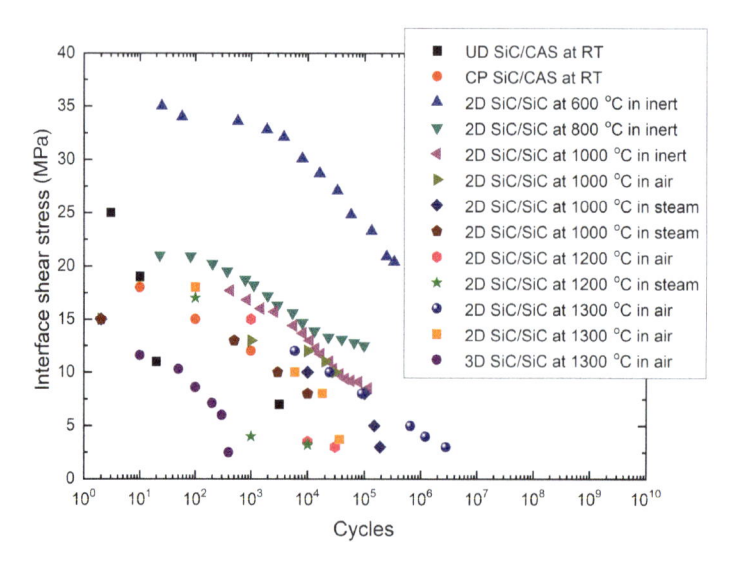

Figure 4.25 The interface shear stress versus applied cycle number of unidirectional and cross-ply SiC/CAS, 2D and 3D SiC/SiC composites at room and elevated temperatures.

For 2D SiC/SiC composite at 1000, 1200, and 1300 °C in air condition, the interface shear stress degradation rate is the highest corresponding to the test condition at 1300 °C in air condition, i.e. the interface shear stress decreases 79.4% from the 100th cycle to the 36 000th cycle, and the lowest for the test condition at 1000 °C in air condition, i.e. the interface shear stress decreases 33.3% from the 2nd cycle to the 30 000th cycle.

For 2D SiC/SiC composite at 1200 °C in air and steam conditions, the interface shear stress degradation rate is higher corresponding to the test condition in steam condition, i.e. the interface shear stress decreases 81.2% from the 100th cycle to the 10 000th cycle, and lower for the test condition in air condition, i.e. the interface shear stress decreases 80% from the 1000th cycle to the 30 000th cycle.

For 2D and 3D SiC/SiC composites at 1300 °C in air condition, the interface shear stress degradation rate is higher corresponding to 3D braided SiC/SiC composite, i.e. the interface shear stress decreases 78.4% from the 10th cycle to the 400th cycle, and lower for 2D woven SiC/SiC composite, i.e. the interface shear stress decreases 79.4% from the 100th cycle to 36 000th cycle.

4.6.2 Comparisons of Interface Damage Between C/SiC and SiC/SiC Composites

The interface shear stress versus cycle number curves of unidirectional C/SiC composite at room temperature and 800 °C in air condition, cross-ply C/SiC composite at room temperature and 800 °C in air condition, 2D C/SiC composite at room temperature, 2.5D C/SiC composite at room temperature, 600 °C in inert, and 800 °C in air, 2D SiC/SiC composite at 600, 800, and 1000 °C in inert, 1000 °C in air and steam, 1200 °C in air and steam, and 1300 °C in air, and 3D SiC/SiC composite at 1300 °C in air, are shown in Figure 4.26. The degradation rate ψ of interface shear stresses in C/SiC and SiC/SiC composite at room and elevated temperatures are listed in Table 4.14.

$$\psi = \frac{\tau_i(N_{\text{initial}}) - \tau_i(N_{\text{final}})}{N_{\text{final}} - N_{\text{initial}}} \tag{4.2}$$

where N_{initial} and N_{final} denote the initial and final cycle number for estimating interface shear stress; and $\tau_i(N_{\text{initial}})$ and $\tau_i(N_{\text{final}})$ denote the estimated interface shear stress at the initial and final cycle number.

For unidirectional, cross-ply, and 2.5D C/SiC at 800 °C in air, and 2D SiC/SiC at 800 °C in inert condition, the interface shear stress degradation rates are 3.4×10^{-4}, 7.7×10^{-4}, 1.8×10^{-4}, and 8.6×10^{-5} MPa/cycle, respectively, as shown in Table 4.14. The interface shear stress degradation rate is much higher at 800 °C in air condition than that at the same elevated temperature in inert condition, due to interface oxidation. For C/SiC composite, the axial thermal expansion coefficient of the carbon fiber ($\alpha_{lf} = -0.38 \times 10^{-6}$ °C) is lower than that of the SiC matrix ($\alpha_{lm} = 2.8 \times 10^{-6}$ °C), the microcracks would appear when the materials cooled down from the high fabricated temperature to room temperature. The radial thermal expansion coefficient of the fiber ($\alpha_{rf} = 7.0 \times 10^{-6}$ °C) is higher than that of the matrix ($\alpha_{rm} = 4.6 \times 10^{-6}$ °C), the thermal residual tensile

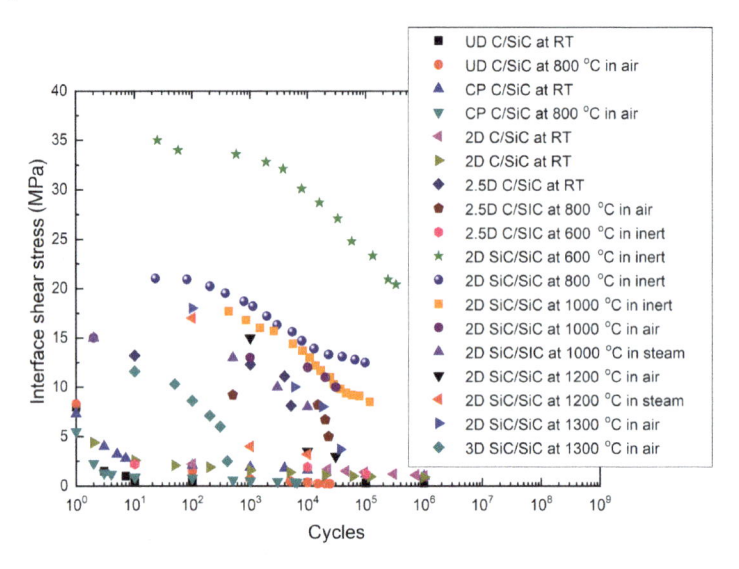

Figure 4.26 The interface shear stress versus cycle number curves of C/SiC and SiC/SiC composite at room and elevated temperatures.

Table 4.14 The interface shear stress degradation rate of C/SiC and SiC/SiC composites at room and elevated temperatures.

	Items	$\tau_{initial}$ (MPa)	τ_{final} (MPa)	$N_{initial}$	N_{final}	ψ (MPa/cycle)
UD C/SiC	RT	8	0.3	1	1 000 000	7.7×10^{-6}
	800 °C in air	8.3	0.2	1	24 000	3.4×10^{-4}
CP C/SiC	RT	7.3	1	1	1 000 000	6.3×10^{-6}
	800 °C in air	5.5	0.4	1	6 600	7.7×10^{-4}
2D C/SiC	RT	2.2	1	100	1 000 000	1.2×10^{-6}
	RT	4.4	0.9	2	1 000 000	3.5×10^{-6}
2.5D C/SiC	RT	13.2	8.1	10	5 210	9.8×10^{-4}
	600 °C in inert	2.2	1.2	10	100 000	1×10^{-5}
	800 °C in air	9.2	5	500	22 700	1.8×10^{-4}
2D SiC/SiC	600 °C in inert	35	20.4	25	333 507	4.4×10^{-5}
	800 °C in inert	21	12.5	23	97 894	8.6×10^{-5}
	1000 °C in inert	17.7	8.5	425	117 055	7.9×10^{-5}
	1000 °C in air	15	10	2	30 000	1.6×10^{-4}
	1000 °C in steam	15	8	2	10 000	7×10^{-4}
	1200 °C in air	15	3	100	30 000	4.1×10^{-4}
	1200 °C in steam	17	3.2	100	10 000	1.4×10^{-4}
	1300 °C in air	18	3.7	100	36 000	3.9×10^{-4}
3D SiC/SiC	1300 °C in air	11.6	2.5	10	400	2.3×10^{-2}

stress would exist in the fiber/matrix interface, and the interface shear stress would be low when interface debonding occurs. At 800 °C in air condition, these matrix microcracks and debonded interface would serve as avenues for the ingress of environment atmosphere into the composite, leading to the oxidation of interphase and then the degradation of interface shear stress. The degradation rate of interface shear stress is the highest for cross-ply C/SiC at 800 °C in air, and the lowest for 2D SiC/SiC at 800 °C in inert.

For 2D C/SiC at room temperature, and 2D SiC/SiC at 600, 800, and 1000 °C in inert, the interface shear stress degradation rates are 1.2×10^{-6}, 4.4×10^{-5}, 8.6×10^{-5}, and 7.9×10^{-5} MPa/cycle, respectively, as shown in Table 4.14. For SiC/SiC composite, the radial thermal expansion coefficient of the SiC fiber ($\alpha_{rf} = 2.9 \times 10^{-6}$ °C) is lower than that of the SiC matrix ($\alpha_{rm} = 4.6 \times 10^{-6}$ °C), the thermal residual compressive stress would exist when the materials cooled down from the high fabricated temperature to room temperature. When interface debonding occurs during fatigue loading, the interface shear stress would decrease with the increasing test temperature.

$$\tau_{imin} = \tau_{LR} + \mu \frac{|\alpha_{rf} - \alpha_{rm}|(T_0 - T)}{A} \tag{4.3}$$

where τ_{LR} denotes the effects of all of the long-range interactions rather than the radial thermal expansion effect; μ denotes the interface frictional coefficient; T denotes the experimental temperature; and T_0 denotes the processing temperature. It can be found in Eq. (4.2) that the interface shear stress would decrease as the test temperature increases, and the interface shear stress degradation rate would increase as the test temperature increases. The degradation rate of interface shear stress is the highest for 2D SiC/SiC at 800 °C in inert, and the lowest for 2D C/SiC composite at room temperature.

For 2.5D C/SiC and 2D SiC/SiC at 600 °C in inert, the interface shear stress degradation rates are 1.0×10^{-5} and 4.4×10^{-5} MPa/cycle, respectively, as shown in Table 4.14. The degradation rate of interface shear stress is higher for 2D SiC/SiC than that of 2.5D C/SiC, due to interface radial thermal residual compressive stress decreasing with increasing test temperature for SiC/SiC composite.

For 2.5D C/SiC at 800 °C in air, 2D SiC/SiC at 1000, 1200, and 1300 °C in air, and 3D SiC/SiC at 1300 °C in air, the interface shear stress degradation rates are 1.8×10^{-4}, 1.6×10^{-4}, 4.1×10^{-4}, 3.9×10^{-4}, and 2.3×10^{-2} MPa/cycle, respectively, as shown in Table 4.14. The degradation rate of interface shear stress is the highest for 3D SiC/SiC at 1300 °C in air, and slowest for 2D SiC/SiC at 1000 °C in air. The interface shear stress degradation rate increases with test temperature in air condition, due to interface oxidation.

4.7 Conclusion

In this chapter, the hysteresis dissipated energy for the strain energy lost per volume during corresponding cycle is formulated in terms of interface shear stress. Comparing experimental fatigue hysteresis dissipated energy with theoretical computational values, the interface shear stress of unidirectional, cross-ply, 2D, and 3D CMCs at room temperature, 600, 800, 1000, 1200, and 1300 °C in inert, air, and steam conditions, are obtained. The effects of test temperature, oxidation, and fiber preforms on the degradation rate of interface shear stress are investigated, and the comparisons of interface degradation between C/SiC and SiC/SiC composites are analyzed.

(1) The interface shear stress decreases with increasing test temperature, and the interface shear stress degradation rate increases with increasing test temperature corresponding to 2D SiC/SiC composite at 600, 800, and 1000 °C in inert atmosphere. For unidirectional, cross-ply, and 2.5D C/SiC at 800 °C in air, 2D SiC/SiC at 800 °C in inert, the degradation rate of interface shear stress is the highest for cross-ply C/SiC at 800 °C in air, and the lowest for 2D SiC/SiC composite at 800 °C in inert condition.
(2) The interface shear stress degradation rate is the highest corresponding to the steam condition, and the lowest for the inert condition, corresponding to 2D SiC/SiC composite at 1000 °C in inert, air, and steam conditions.
(3) The interface shear stress degradation rate is the highest corresponding to the test condition at 1300 °C in air, and the lowest for the test condition at 1000 °C in air, corresponding to 2D SiC/SiC composite at 1000, 1200, and 1300 °C in air condition.
(4) For 2.5D C/SiC at 800 °C in air, 2D SiC/SiC at 1000, 1200, and 1300 °C in air, and 3D SiC/SiC at 1300 °C in air, the degradation rate of interface shear stress is the highest for 3D SiC/SiC at 1300 °C in air, and the slowest for 2D SiC/SiC at 1000 °C in air condition.

References

Bednarcyk, B.A., Mital, S.K., Pineda, E.J., and Arnold, S.M. (2015). Multiscale modeling of ceramic matrix composites. *56th AIAA/ASCE/AHS/ASC Structures, Structural Dynamics, and Materials Conference*, Kissimmee, Florida (5–9 January 2015). Reston, VA, USA: AIAA (American Institute of Aeronautics and Astronautics). https://doi.org/10.2514/6.2015-1191.

Brandstetter, J., Kromp, K., Peterlik, H., and Weiss, R. (2005). Effect of surface roughness on friction in fiber-bundle pull-out tests. *Composites Science and Technology* 65: 981–988. https://doi.org/10.1016/j.compscitech.2004.11.004.

Chandra, N. and Ghonem, H. (2001). Interfacial mechanics of push-out tests: theory and experiments. *Composites Part A Applied Science and Manufacturing* 32: 575–584. https://doi.org/10.1016/S1359-835X(00)00051-8.

Cho, C.D., Holmes, J.W., and Barber, J.R. (1991). Estimation of interfacial shear in ceramic composites from frictional heating measurements. *Journal of the American Ceramic Society* 74: 2802–2808. https://doi.org/10.1111/j.1151-2916.1991.tb06846.x.

Curtin, W.A. (2000). Stress–strain behavior of brittle matrix composites. In: *Comprehensive Composite Materials* (editors-in-chief, A. Kelly and C. Zweben), vol. 4, 47–76. Elsevier Science Ltd. https://doi.org/10.1016/B0-08-042993-9/00088-7.

Evans, A.G., Zok, F.W., and McMeeking, R.M. (1995). Fatigue of ceramic matrix composites. *Acta Metallurgica et Materialia* 43: 859–875. https://doi.org/10.1016/0956-7151(94)00304-Z.

Fantozzi, G. and Reynaud, P. (2009). Mechanical hysteresis in ceramic matrix composites. *Materials Science and Engineering A* 521–522: 18–23. https://doi.org/10.1016/j.msea.2008.09.128.

Holmes, J.W. and Cho, C.D. (1992). Experimental observations of frictional heating in fiber-reinforced ceramics. *Journal of the American Ceramic Society* 75: 929–938. https://doi.org/10.1111/j.1151-2916.1992.tb04162.x.

Jacob, D. (2010). Fatigue behavior of an advanced SiC/SiC composite with an oxidation inhibited matrix at 1200 °C in air and in steam. Master thesis. Air Force Institute of Technology, Ohio, USA.

Kim, J. and Liaw, P.K. (2005). Characterization of fatigue damage modes in nicalon/calcium aluminosilicate composites. *Journal of Engineering Materials and Technology* 127: 8–15.

Kuntz, M. and Grathwahl, G. (2001). Advanced evaluation of push-in data for the assessment of fiber reinforced ceramic matrix composites. *Advanced Engineering Materials* 3: 371–379. https://doi.org/10.1002/1527-2648(200106)3:6<371::AID-ADEM371>3.0.CO;2-Y.

Li, L. (2014). Assessment of the interfacial properties from fatigue hysteresis loss energy in ceramic-matrix composites with different fiber preforms at room and elevated temperatures. *Materials Science and Engineering A* 613: 17–36. https://doi.org/10.1016/j.msea.2014.06.092.

Li, L. (2017). Comparisons of interface shear stress degradation rate between C/SiC and SiC/SiC ceramic-matrix composites under cyclic fatigue loading at room and elevated temperatures. *Composite Interfaces* 24: 171–202. https://doi.org/10.1080/09276440.2016.1196995.

Li, L. and Song, Y. (2010). An approach to estimate interface shear stress of ceramic matrix composites from hysteresis loops. *Applied Composite Materials* 17: 309–328. https://doi.org/10.1007/s10443-009-9122-6.

Li, L., Song, Y., and Sun, Y. (2013). Estimate interface shear stress of unidirectional C/SiC ceramic matrix composites from hysteresis loops. *Applied Composite Materials* 20: 693–707. https://doi.org/10.1007/s10443-012-9297-0.

Liu, C.D., Cheng, L.F., Luan, X.G. et al. (2008). Damage evolution and real-time non-destructive evaluation of 2D carbon-fiber/SiC-matrix composites under fatigue loading. *Materials Letters* 62: 3922–3924. https://doi.org/10.1016/j.matlet.2008.04.063.

Mall, S. and Engesser, J.M. (2006). Effects of frequency on fatigue behavior of CVI C/SiC at elevated temperature. *Composites Science and Technology* 66: 863–874. https://doi.org/10.1016/j.compscitech.2005.06.020.

Michael, K. (2010). Fatigue behavior of a SiC/SiC composite at 1000 °C in air and steam. Master thesis. Air Force Institute of Technology, Ohio, USA.

Moevus, M., Reynaud, P., R'Mili, M. et al. (2006). Static fatigue of a 2.5D SiC/[Si–B–C] composite at intermediate temperature under air. *Advances in*

Science and Technology 50: 141–146. https://doi.org/10.4028/www.scientific.net/AST.50.141.

Naslain, R. (2004). Design, preparation and properties of non-oxide CMCs for application in engines and nuclear reactors: an overview. *Composites Science and Technology* 64: 155–170. https://doi.org/10.1016/S0266-3538(03)00230-6.

Opalski, F.A. and Mall, S. (1994). Tension-compression fatigue behavior of a silicon carbide calcium-aluminosilicate ceramic matrix composites. *Journal of Reinforced Plastics and Composites* 13: 420–438. https://doi.org/10.1177/073168449401300503.

Reynaud, P. (1996). Cyclic fatigue of ceramic-matrix composites at ambient and elevated temperatures. *Composites Science and Technology* 56: 809–814. https://doi.org/10.1016/0266-3538(96)00025-5.

Rouby, D. and Louet, N. (2002). The frictional interface: a tribological approach of thermal misfit, surface roughness and sliding velocity effects. *Composites Part A Applied Science and Manufacturing* 33: 1453–1459. https://doi.org/10.1016/S1359-835X(02)00145-8.

Rouby, D. and Reynaud, P. (1993). Fatigue behavior related to interface modification during load cycling in ceramic-matrix fiber composites. *Composites Science and Technology* 48: 109–118. https://doi.org/10.1016/0266-3538(93)90126-2.

Shi, D., Jing, X., and Yang, X. (2015). Low cycle fatigue behavior of a 3D braided KD-I fiber reinforced ceramic matrix composite for coated and uncoated specimens at 1100 °C and 1300 °C. *Materials Science and Engineering A* 631: 38–44. https://doi.org/10.1016/j.msea.2015.01.078.

Vagaggini, E., Domergue, J.M., and Evans, A.G. (1995). Relationships between hysteresis measurements and the constituent properties of ceramic matrix composites: I, Theory. *Journal of the American Ceramic Society* 78: 2709–2720. https://doi.org/10.1111/j.1151-2916.1995.tb08046.x.

Yang, C.P., Jiao, G.Q., Wang, B., and Du, L. (2009). Oxidation damages and a stiffness model for 2D-C/SiC composites. *Acta Materiae Compositae Sinica* 26: 175–181.

Zhu, S.J., Mizuno, M., Kagawa, Y., and Mutoh, Y. (1999). Monotonic tension, fatigue and creep behavior of SiC-fiber-reinforced SiC-matrix composites: a review. *Composites Science and Technology* 59: 833–851. https://doi.org/10.1016/S0266-3538(99)00014-7.

Zhu, S.J., Mizuno, M., Nagano, Y. et al. (1998). Creep and fatigue behavior in an enhanced SiC/SiC composite at high temperature. *Journal of the American Ceramic Society* 81: 2269–2277. https://doi.org/10.1111/j.1151-2916.1998.tb02621.x.

Index

a
aero-engine core engines   29
α-Al_2O_3 15
alumina matrix oxide/oxide composites   15

b
BN interphase oxidation   8

c
ceramic-matrix composites (CMCs)
  aero-engine core engines   29
  BN interphase   21
  characteristics of   29
  combustion chamber and high-pressure turbine components   161
  cross-ply CMCs   122–124
  cross-ply C/SiC composite   131–135, 144–155
  C/SiC and SiC/SiC composites   193–195
  cyclic loading/unloading/fatigue hysteresis behavior of   110
  at elevated temperature in air atmosphere
    1000°C   172–175
    1200°C   175–178
    1300°C   178–184
  at elevated temperature in intert atmosphere   167–172
  at elevated temperature in steam atmosphere
    1000°C   185–187
    1200°C   187–191

  experimental investigations
    fatigue hysteresis-based damage parameters   46–51
    first matrix cracking stress   42–43
    matrix cracking density   43–46
  fiber/matrix interface shear stress   30, 110, 111
  fiber-reinforced CMCs   29
  first matrix cracking stress   31
  fracture modes of monolithic ceramics   1
  high quality and high temperature properties   29
  hysteresis theories   112–118
  interface damage law
    at room temperature   163–167
  interface effects
    sliding resistance   19
    thermal misfit stress   19
  interface properties, effects of
    fatigue hysteresis-based damage parameters   32–33, 39–41
    first matrix cracking stress   31, 33–36
    matrix cracking density   31–32, 36–38
    theoretical analysis   31–33
  life prediction model
    at elevated temperature   89–100
    at elevated temperatures in oxidative environment   77–79
    experimental investigations   79–89
    at room temperature   74–77, 79–89

ceramic-matrix composites (CMCs) (*contd.*)
 theoretical analysis   73–79
 matrix cracking density   32
 monolithic ceramic and fiber-reinforced   1, 2
 non-oxide CMCs   3–13
 oxide/oxide composites
  $\alpha\text{–}Al_2O_3$ 15
  Al_2O_3 and $Al_2O_3\text{-}SiO_2$ ceramic fibers   14
  environmental stability   13–17
  fracture behavior of   15
  fugitive coating   16
  $LaPO_4$ 15
  matrix materials   15
  PIP 8HSW Nextel 720/alumina composite   16
  properties of   14
  2D Nextel 610/ $Al_2O_3\text{-}SiO_2$ composite   16
  2D Nextel 720TM/alumina composite   17
 PyC interphase   20–21
 research and application of   29
 shape, location and area of hysteresis loops   162
 stress-strain hysteresis loop   109
 temperature, oxidation, and fiber preforms   192–193
 tensile damage and fracture process
  experimental investigations   66–71
  fiber failure, interface and fiber oxidation   58–59
  fiber strength, effect of   64–65, 67
  fiber Weibull modulus, effect of   65–66
  interface debonding   57, 58
  interface shear stress, effects of   62–64
  matrix multicracking   56–57
  pre-exposure temperature   60–61
  pre-exposure time, effects of   61–62
  stress analysis   54–56
  tensile stress-strain curves   59–60
 theoretical analysis   52–60
 2.5D C/SiC composite   135–138, 149–155
 2D CMCs   124–127
 unidirectional CMCs   118–123
 unidirectional C/SiC composite   129–131, 138, 140–142
 unidirectional SiC/CAS and SiC/SiC composites   30
 unidirectional SiC/CAS composite   142–144
 unidirectional SiC/CAS-II composite   144, 145
cooling methods   1
cross-ply ceramic composite   131
cross-Ply C/SiC composite   131, 133–136
CVI Hi-Nicalon SiC/SiC composite   11

e
Embrittled fibers   8
energy-disspipating mechanism   110
Evans–Zok–McMeeking model   80, 84, 89, 93, 96, 100
experimental fatigue hysteresis dissipated energy   174, 179

f
fatigue hysteresis behavior   129
fatigue hysteresis dissipated energy   139, 140
 versus cycle number curves   168
 of cross-ply SiC/CAS composite   167
 of 2D SiC/SiC composite   171, 173, 175
 of unidirectional SiC/CAS composite   165
fatigue hysteresis loops   119, 121, 123, 125, 129, 131–134, 136, 137
fatigue stress ratio   11, 122, 142, 149, 151
Fatigue tests   11, 13, 16, 17, 172, 175, 178, 180, 185, 187
fiber axial stress distribution   113–118

f

fiber/matrix interface shear stress
  29–31, 42, 54, 110, 127, 161, 162
fiber-reinforced CMCs 1, 2, 21, 29–31,
  51, 59, 71, 79, 100, 112, 155
fiber Weibull modulus 31, 59, 60, 65,
  66, 68
F414 turbofan 161

g

gas turbine efficiency 1
GEnx aero engine 161
Global Load Sharing criterion 58

h

Hutchinson–Jensen fiber pull-out model
  110
hysteresis dissipated energy
  vs. interface shear stress diagram
  164

i

infrared pyrometers 118
interface shear stress 84
  vs. cycle number curve 165, 167,
  171–173, 175
interface slip mechanism 110
interface wear process 111, 161

l

$LaPO_4$ 15

m

matrix cracking density 31–32,
  36–39, 43–46, 57, 100, 103
matrix Weibull modulus 57
micromechanics approach 30, 72, 100
MI Hi-Nicalon SiC/SiC composite 4,
  7, 11, 12
Mullite ($3Al_2O_3$-$2SiO_2$) 15

n

Nextel 312 14
Nextel 610 14–16
Nextel 610/Al_2O_3 composite 15, 16
Nextel 550 fiber 14
Non-oxide CMCs 2–13, 21

o

oxidation embrittlement 3, 13

p

PIP 8HSW Nextel 720/alumina
  composite 16
PIP SiC/SiC composite 11, 12
prepreg-MI SiC/SiC composites 4, 6
Pryce–Smith model 110
PyC interphace ression 162
PyC interphase 9–11, 19–22, 93, 96,
  100, 103, 111, 162

s

sapphire/$LaPO_4$/alumina composite
  system 15
shear-lag model 54, 56, 114
SiC/PyC/borosilicate glass composite
  19
SiC/SiC composites
  with BN interphase at room
  temperature 6
  degradation behavior of 13
  with different fiber preforms 7
  oxidation heat-treatment of 8
  radial stress 9
stress-strain hysteresis loops 9, 100,
  110, 112, 118, 120, 123, 125

t

theoretical fatigue hysteresis dissipated
  energy
  vs. interface shear stress curve of 2D
  SiC/SiC composite 174, 179
thermal misfit stress 19
thermal residual stress (TRS) 19, 57
3D C/SiC composite 66, 67, 76, 89, 92,
  100, 102, 110
3D tyranno SiC/[Si-Ti-C-O] composite
  11
2D C/SiC composite 68, 80, 84, 93, 96,
  109–112, 155, 162, 193, 195
2D CVI Nicalon SiC/SiC composite 9,
  10

2D CVI SiC/SiC composite
  oxidation of fatigue beahvior of 11
  tensile and fatigue behaviour 10
2D CVI Sylramic-iBN/BN/SiC
  composite 12, 13
2D Hi-Nicalon SiC/SiC composite 4,
  7, 11, 12, 42, 49, 54
2D MI Sylramic-iBN/BN/SiC composite
  12, 13
2D Nextel 610/Al_2O_3-SiO_2 composite
  16
2D Nextel 720-alumina composite 17
2D SiC/SiC composite 167, 187
  fatigue hysteresis loops area of 109
2D Sylramic-SiC/SiC composite 46
2D Tyranno-SiC/SiC composite 46, 50
2.5D C/SiC composite 89, 96

two-parameter Weibull distribution
  56, 72
two-parameter Weibull model 73

u

unidirectional C/SiC composite 60,
  70, 79–81, 129, 131, 138–142,
  155
unidirectional SiC/CAS-II composite
  118, 119, 121, 123, 144, 145,
  162

v

Vagaggini's hysteresis loops models
  111, 162
Vagaggini's unidirectional hysteresis
  loops models 110